PE
1475
.M37
1986

Master, Peter
   Antony.

Science, medicine,
  and technology

Cop. 2

14.95

JUL 1987	DATE		

S-2

SCIENCE, MEDICINE, and TECHNOLOGY:
English Grammar and Technical Writing

PETER ANTONY MASTER

University of California at Los Angeles

PRENTICE-HALL, INC., Englewood Cliffs, New Jersey 07632

Library of Congress Cataloging-in-Publication Data

MASTER, PETER ANTONY.
   Science, medicine, and technology.

   Includes bibliographies and index.
     1. English language—Technical English.   2. English
language—Rhetoric.   3. English language—Text-books
for foreign speakers.   4. English language—Grammar—
[date].   5. Technical writing.   I. Title.
[DNLM: 1. Linguistics.   2. Medicine.   3. Science.
4. Technology.   5. Writing.   T 11 M423s]
PE1475.M37   1986   808'.0666   85-16918
ISBN 0-13-795469-7

Editorial/production supervision and
   interior design: *Edith Riker*
Cover design: *20/20 Services, Inc.*
Manufacturing buyer: *Harry Baisley*
Illustrations: *Angela Candela Morley*

Printed in the United States of America

10 9 8 7 6 5 4 3 2 1

ISBN 0-13-795469-7   01

Prentice-Hall International (UK) Limited, *London*
Prentice-Hall of Australia Pty. Limited, *Sydney*
Prentice-Hall Canada Inc., *Toronto*
Prentice-Hall Hispanoamericana, S.A., *Mexico*
Prentice-Hall of India Private Limited, *New Delhi*
Prentice-Hall of Japan, Inc., *Tokyo*
Prentice-Hall of Southeast Asia Pte. Ltd., *Singapore*
Editora Prentice-Hall do Brasil, Ltda., *Rio de Janeiro*
Whitehall Books Limited, *Wellington, New Zealand*

WE

FOR MY PARENTS,
ANTONY EDWARD AND EVA JANA MASTER

CONTENTS

PREFACE

Science, Medicine, and Technology: English Grammar and Technical Writing is a textbook for foreign students who are studying or have studied science, medicine, or technology. It is designed to motivate ESL students to learn English, appealing to their sense of what they will need in their professional lives by teaching the common forms of scientific writing and the grammar necessary to produce them. To derive full advantage from this book, students should therefore have knowledge of basic chemistry, biology, mathematics, physics, etc. With such a background, even students with nonscientific and nontechnical majors can benefit from the text.

The book is divided into six units representing seven major rhetorical patterns in technical writing (Unit VI includes two such patterns). Each unit is divided into a grammar section and a writing section.

The grammar section is divided into five major categories: articles, sentence combining, reduction and modification, verbs, and writing aids. Each unit (except Unit I, which has a special section on definitions) repeats these five categories, creating a kind of spiral. As much as possible, the grammar presented in each unit is linked to the rhetorical pattern to which the unit is devoted. For example, defining relative clauses are presented in Unit I because they are needed in constructing a formal definition, the opening sentence of an amplified definition; noun-phrase parallelism is presented in Unit II because it is required in constructing a thesis sentence (or plan-of-development sentence, as I call it) for a description of a mechanism. Of course, many aspects of grammar cannot be linked to rhetorical patterns in this way. These aspects are simply included under the appropriate category heading. Within each category throughout the book, there is either a hierarchy of difficulty (for example, definite articles precede generic articles; the passive precedes infinitive and *that*-complements) or a logical sequence (for instance, relative clauses precede noun compounds; modals precede hedging). Each exercise is written in scientific language and reflects a variety of disciplines. A star (*) before a sentence indicates that it is ungrammatical (incorrect English).

The writing section is also divided into five categories: prewriting activity, structure, models, analysis, and choice of topic. The prewriting activity is concerned with some aspect of organization or problem solving that will be necessary in creating the written assignment at the end of the unit. The structure section breaks down the rhetorical pattern, sentence by sentence, paragraph by paragraph, describing how a typical technical composition is constructed, with appropriate examples. Models are provided from different fields. In the analysis section, the structure is restated in a formulaic manner, and a partner/small-group exercise is provided in which students are asked to analyze the model in terms of the structure described. This exercise demonstrates to students the manner in which the structure was used in creating the model, although certain departures from the structure will be evident too. Finally, a list of topics is provided in various disciplines (all of which have been successfully used by ESL students) to stimulate the students' ideas for their own topics, which should come from their chosen fields of specialization.

TO THE TEACHER

Timing and the use of the text. The grammar and writing sections of each unit should be taught so that both are completed at the same time. The five grammatical categories are provided so that you can easily vary the presentation and discussion of grammar. In other words, as soon as the class becomes tired of working on articles, you can jump to another section—sentence combining or verbs, perhaps—and continue with the articles on another day. This feature would be wasted if you simply worked chronologically through the book, covering first articles, then sentence combining, then modification and reduction, and so on.

Another approach would be to have the students focus on the writing section first and generate a composition. Based on errors in the compositions, you could then direct the class towards the grammatical structures that have caused particular difficulty. The book is constructed on a kind of grid pattern to provide maximum flexibility.

Communicative activities and the writing process. No attempt is made to make the grammar exercises communicative. It has been my experience that many students of science and technology, accustomed as they are to relatively formulaic presentations, find class time spent on internalizing any given structure rather unproductive. Nevertheless, communicative activities can lighten the classroom atmosphere, and attempts to provide an environment wherein certain grammatical structures might be used can be beneficial. This is beyond the scope of this text, however.

Similarly, little attempt is made to encourage the process of writing. Prewriting activities are provided, but brainstorming, the generation of a first draft and subsequent revisions, peer review, conferencing, and so on—all necessary components of the writing process—are left to you. I have merely detailed the structure of a typical rhetorical pattern in technical writing in order to remove the burden of format and allow the student to concentrate on language and organization.

TO THE STUDENT

Science, Medicine, and Technology has also been designed for self-study. For this reason a glossary within the text and an answer key (available separately) have been provided. You should work through the book chronologically so that you can take advantage of the built-in hierarchy of difficulty. Of course, any compositions written will require somebody to correct them, ideally in such a way that grammatical errors are identified for review in the grammar section.

ACKNOWLEDGMENTS

I would like first to acknowledge John Swales for his seminal work *Writing Scientific English*, through which I first came to realize that the English of science and technology could be limited and described.

For their invaluable assistance in editing and proofreading, I am deeply grateful to Ellen Rosenfield and Dr. Jay Reich.

For the illustrations and her patience with my demands, I thank Angela Candela Morley.

For help with the logic and mathematical aspects of the book, I thank Robert Mutti.

And for their crucial comments and their enthusiasm for this project, I thank all my students of the last several years.

UNIT I
The Amplified Definition

PART I
Grammar

Section 1
ARTICLES

The article system in English is one of the most difficult grammatical points for an ESL student to master. Even other languages that have an article system differ from English in their rules for the use of articles.

There are four articles in English:

1. *a*
2. *an*
3. *the*
4. Ø (no article), called the zero article

Every noun in English requires *a, an, the, or* Ø (or another determiner—see Unit III, Section 1), and for every noun we must answer four questions:

1. Is the noun countable or uncountable?
2. Is the noun definite or indefinite?
3. Is the noun generic or specific?
4. Is the noun common or proper?

Question 1 will be discussed in this unit, question 2 in Unit II, question 3 in Unit IV, and question 4 in Unit VI, Section 1.

A VERSUS *AN*

Before we discuss the use of the articles *a* and *an*, try the following exercise.

EXERCISE 1.1   Underline *a* and *an* in the following passage. Try to make a rule about the difference between these two articles.

Time is measured in a variety of ways. A nanosecond is one of the smallest units. An eon is one of the largest. In our daily life we are more interested in time on a human scale: a minute, an hour, a month, a year. An objective measurement of time is necessary for scientific investigation. However, we also recognize subjective time. A one-hour movie can feel like less time than sixty minutes of a boring lecture.

Twice in the passage the same letter follows both *a* and *an:*

a <u>h</u>umanscale   a <u>o</u>ne-hour movie
an <u>h</u>our   an <u>o</u>bjective measurement

Therefore, your rule must be more specific than

a + consonant
an + vowel

A phonetic transcription of these examples

shows that the **sound** of the word following the article controls the choice of *a* or *an:*

a + consonant sound
an + vowel sound

EXERCISE 1.2   Circle the correct answer.

Our solar system, (a an) one-star system of nine planets, was formed approximately 4.5 billion years ago from (a an) universe of cosmic gas and dust. Life in the form of bacterial cells already existed (a an) billion years later. How did life form? Did (a an) X ray from space strike (a an) carbon atom and (a an) hydrogen atom in just the right manner to produce the compound of life? Was it (a an) lightning flash? Or was it (a an) ultraviolet-light source? Nobody has (a an) exact answer.

EXERCISE 1.3   Put *a* or *an* in the blanks.

A STUDENT ADVISER

_____ student adviser at _____ university can help _____ student with _____ variety of problems. _____ undergraduate adviser can help _____ student to decide on _____ major and _____ course of study. If _____ student is _____ honor student, the adviser can recommend _____ scholarship or _____ grant. If _____ student is _____ F student (_____ failing student), the adviser can help with personal problems if necessary and suggest ways to improve study habits. _____ foreign-student adviser is often _____ helpful friend to students who have come to study in _____ new country for the first time.

COUNTABLE AND UNCOUNTABLE NOUNS

It is easy to decide whether most nouns are countable or uncountable. We can count *books* and *elements* and *hypotheses*. We cannot count *wood* and *air* and *chemistry*.

Uncountable nouns can be made countable in several ways. SOLIDS and SPECIFIC INFORMATION have *a piece of* before the noun:

a piece of wood	a piece of information
a piece of cake	a piece of data
a piece of advice	

POWDERS, LIQUIDS, and GASES can be made countable by using more specific words to describe the smallest unit:

a grain of salt	a molecule of water
a particle of sand	a molecule of oxygen
a speck of dust	

Powders, liquids, and gases can also be made countable by containing or limiting them:

a teaspoon of sugar	or	a crystal of sugar
a glass of water		a jet of water
a tank of air		a blast of air

SUBJECTS, LANGUAGES, INTERESTS, and other abstract nouns can be made countable only by changing them to adjective-plus-noun or noun-plus-noun (noun-compound) structures:

	Adjective-plus-Noun	Noun-plus-Noun
SUBJECTS:	a chemical analysis	a chemistry book
	a medical dictionary	a medicine man
LANGUAGES:	a Chinese dinner	the China Sea
INTERESTS:	a photographic plate	a photography course

Noun compounds will be discussed in Unit VI, Section 3.

EXERCISE 1.4 Circle the word on the right that is the best uncountable synonym for the countable word on the left.

Example: a banana → food fruit nutrition
Answer: a banana → food (fruit) nutrition

1. a table → furniture   wood   matter
2. a thermometer → measurement   instrumentation   equipment
3. a piece of data → knowledge   information   research
4. a dollar → change   money   capital
5. a book → literature   information   research
6. a hammer → force   equipment   steel
7. a train → machinery   freight   transportation
8. a sentence → information   language   linguistics
9. a computer program → data   software   information
10. a sugar molecule → sweetness   sugar   energy

EXERCISE 1.5 Make a list of the types of uncountable nouns described above (solids, specific information, liquids, etc.). Think of three examples of each class. How would you make your examples countable?

Example: solids

wood → a piece of wood, a log
ice → a piece of ice, an ice cube
iron → a piece of iron, a nail

MASS NOUNS

There is a last group of uncountable nouns, and these may appear to be countable. Consider the word *money*. You can count dollars or yen or pesos or coins or bills, but not money. The word *money* belongs to the type of uncountable nouns called mass nouns—nouns that represent groups of countable nouns.

Mass Noun (uncountable)	Individual examples (countable)
money	dollars, pounds, yen, pesos, coins, nickels
furniture	chairs, tables, beds, lamps
fruit	apples, bananas, papayas
equipment	thermometers, scalpels, test tubes
clothing	shirts, dresses, pants, suits
machinery	engines, bulldozers, sewing machines
news	stories, reports, bulletins, events

Mass nouns are often preceded by the determiner *some*, and they may be made countable through the use of *a piece of*:

> We bought <u>some</u> new equipment for the lab.
> The newest <u>piece of</u> equipment was bought in 1946.

EXERCISE 1.6   Write a countable form of the following uncountable words or phrases.

Example: water _____
 Answer: water   <u>a glass of water, a liquid</u>

1. paper _____
2. homework _____
3. traffic _____
4. information _____
5. NaOH (sodium hydroxide) _____
6. blood _____
7. vegetation _____
8. electricity _____
9. heat _____
10. oxygen _____

DUAL NOUNS

There are many nouns in English that may be both countable and uncountable. These are called dual nouns. Such nouns may have a completely different meaning in each case:

iron (Fe, an element)   an iron (an instrument for smoothing clothes)
glass (a clear, hard   a glass (a container for a potable liquid)
  silicate)

EXERCISE 1.7   Determine the difference between the countable and un-countable forms of the words *wood, paper, light,* and *man.* Write two sentences with each word, one containing the countable form and the other the uncount-able form.

Example: Steel is made from iron.
I burned my finger on the iron.

Scientists often use the countable/uncountable difference to indicate a general sense versus a specific one:

Grass is eaten by cows and other grazing animals.
(In this sentence, *grass* means *grass in general.* We are not interested here in different types of grass.)
Some grasses stay green all year; others turn yellow in the summer.
(In this sentence, *grasses* means *different kinds of grass.* We are interested here in specific types.)

EXERCISE 1.8   Indicate whether the underlined word is specific (S) or gen-eral (G). Then indicate whether it is countable (C) or uncountable (U)?

Example: _____ Paper is made from wood. _____
Answer: __G__ Paper is made from wood. __U__

S/G   C/U

1. _____ Weight is a function of gravity.   _____
2. _____ They measured a weight of forty-five grams.   _____
3. _____ A high-carbon steel offers more resistance to
corrosion.   _____
4. _____ Steel is manufactured in the eastern U.S.   _____
5. _____ Some grains cannot grow in cool climates.   _____
6. _____ Grain is used, among other things, to feed
livestock.   _____
7. _____ Stone was the first material used for tools.   _____
8. _____ Stones were the first objects used for tools.   _____
9. _____ The patient had an infection in her eye.   _____
10. _____ Infection is usually caused by bacteria.   _____

EXERCISE 1.9   Cross out the incorrect form of the dual noun in paren-theses. Is the form countable or uncountable?

Example: He drank (glass/a glass) of water.   _____
Answer: He drank (a glass) of water.   countable

1. The geologist found (diamond/a diamond) in his pocket. _____
2. A tire is made of (rubber/a rubber). _____

3. The sun provides us with (light/a light). ———————
4. (Rope/A rope) is imported from the Philippines. ———————
5. There was (fire/a fire) in the apartment downstairs. ———————
6. A living heart has (sound/a sound) like a drum. ———————
7. (Paper/A paper) is manufactured from wood pulp. ———————
8. The laboratory did (analysis/an analysis) of the blood sample. ——
9. (Football/A football) is popular all over the world. ———————
10. To complete the circuit, connect (wire/a wire) to the battery. ——

EXERCISE 1.10   For each of the sentences in exercise 1.9, write the correct article (*a* or *Ø*) in the blank provided below. Indicate whether the dual noun is specific or general by circling the appropriate word. Then write a definition of the noun.

Example: _ glass   specific   general
 Answer: a glass (specific) general
Definition: A glass is a drink container that is made of glass.

1. ———— diamond   specific   general
   Definition: ———————————————————————
2. ———— rubber   specific   general
   Definition: ———————————————————————
3. ———— light   specific   general
   Definition: ———————————————————————
4. ———— rope   specific   general
   Definition: ———————————————————————
5. ———— fire   specific   general
   Definition: ———————————————————————
6. ———— sound   specific   general
   Definition: ———————————————————————
7. ———— paper   specific   general
   Definition: ———————————————————————
8. ———— analysis   specific   general
   Definition: ———————————————————————
9. ———— football   specific   general
   Definition: ———————————————————————
10. ———— wire   specific   general
   Definition: ———————————————————————

A(N) VERSUS ONE

In some cases, the article *a(n)* appears to be the same as the word *one:*

Our class meets five times a week (five times in one week).
He lost a hundred dollars (lost one hundred dollars).
She earns $25,000 a year ($25,000 for one year).

However, *a(n)* is not identical to *one*, as it is in many other languages. Can you determine the difference between *a(n)* and *one* from the following examples?

He used <u>a</u> pipette. =   He used an instrument that is called a pipette.
He used <u>one</u> pipette. = He only used one pipette; he didn't use two or three.

A(n) refers to any single item. *One* refers to a limited number.

EXERCISE 1.11   Write *a(n)* and/or *one* in the blank. If both answers are possible, explain the difference in meaning.

1. The surgeon needs _____ scalpel.
2. A Bunsen burner is _____ excellent source of heat in the laboratory.
3. Please read _____ chapter _____ week.
4. You have many good ideas, but we must come up with _____ design for this factory.
5. An egg accepts _____ sperm and rejects all others.
6. A mayfly has an average life-span of _____ day.
7. The steam engine is _____ important source of power.
8. A banana tree produces _____ bunch of bananas and then dies.
9. An engineer usually earns at least $35,000 _____ year.
10. Each element has _____ characteristic spectrum.

EXERCISE 1.12   Fill the blanks with *a, an,* or *one.* If there is more than one possible answer, how does the meaning change?

Because there are many fascinating fields in science and technology, it is sometimes difficult to decide on _____ major. Should _____ student study _____ subject that is really interesting, or should the student study _____ subject that will pay _____ high salary? What if the student is interested not just in _____ subject but in two or three? _____ solution is to select _____ combined major (for example, biology and engineering). Many people believe that if _____ student studies for _____ reason only—money— his or her career will not be as successful.

A(N) VERSUS Ø (THE ZERO ARTICLE)

In deciding whether to use *a(n)* or *Ø* remember that *a(n)* is related (but not identical) to *one*.

a book	benzene	opinions	a microprocessor
an idea	a glass	quality	strength
books	research	a solid	gases

Can you make a rule for the use of *a(n)* and *Ø* from the above list of words?

We have said that *a(n)* means **any** single item. For this reason, *a(n)* is called an indefinite article. It is impossible to use *a(n)* with an uncountable noun because an uncountable noun is a quantity, not a single item. It is impossible to use *a(n)* with plural countable nouns because *plural* means *more than one*. For plural and uncountable nouns, we must therefore use the zero article. The rule for the use of the indefinite articles is:

Article	Type of Noun	Example
a(n)	+ singular countable	a book
Ø	+ uncountable	air
Ø	+ plural countable	books

You have probably noticed that an article always comes **before** an adjective or postdeterminer (see Unit III, Section 1, "Articles With Premodified Noun Phrases"):

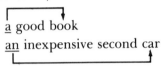

a good book

an inexpensive second car

EXERCISE 1.13   Write *a(n)* or Ø in the blanks.

HOSPITALS

_____ hospital is _____ place for the scientific treatment of _____ sick people. _____ many modern hospitals are also _____ medical centers where _____ doctor can send _____ patient for _____ examination and _____ diagnosis as well as for _____ treatment. _____ some hospitals are responsible for training _____ doctors, _____ nurses, and _____ other medical personnel. _____ hospitals are also _____ research centers where _____ new drug or _____ special surgical procedures and _____ treatments are developed.

EXERCISE 1.14   With another student, underline the head nouns (the main nouns, not the modifying ones) in the following passage. Choose the correct indefinite article, (*a(n)* or Ø), and put it in the correct place. See Figure 1.1 on page 10.

Good laboratory is essential for good research. It should have strong table with hard, acid-resistant surface. It should have water faucet and sink with controlled drainage. It should have gas outlet for Bunsen burner and stands for holding test tubes, flasks, and other laboratory equipment. Well-equipped laboratories also supply hot steam, vacuum, and possibly oxygen or other gases.

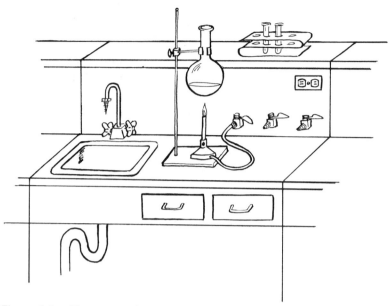

Figure 1.1 Chemistry Laboratory Desk

Section 2
DEFINITIONS

THE FORMAL DEFINITION

It is often necessary in scientific writing to explain what a certain word means. The best way to do this is to use Aristotle's well-known definition formula

An $\boxed{A}$ is a $\boxed{B}$ that $\boxed{C}$ $(+C_1+C_2+C_3,$ etc.)

$\boxed{\text{A thermometer}}$ is $\boxed{\text{an instrument}}$ that $\boxed{\text{measures temperature.}}$

$\boxed{\text{A vitamin}}$ is a $\boxed{\text{complex organic substance}}$

that $\boxed{\text{a plant or animal must obtain from its environment}}$

and that $\boxed{\text{always plays an essential part in its metabolism.}}$

You can see that

A = the species (the word or words we want to define)

B = the general group or class that includes A

C = the characteristic(s) that make(s) A different from other examples of B

EXERCISE 2.1   Determine which words belong to A (species) and which to B (general class). Then match each A with an appropriate B (the B words may be used more than once).

Example: <u>A</u> chair → <u>B</u> a piece of furniture

__ benzene	__ an element	__ a compound	__ steam
__ a cow	__ equipment	__ a farm animal	__ rust
__ a nanosecond	__ oxygen	__ a Bunsen burner	__ an infrared
__ a type of	__ an X ray	__ carbon	wave
corrosion	__ a test tube	__ an eon	__ a vitamin
__ silver	__ a sheep	__ a unit of time	__ an electromag-
sulphide	__ a pipette	__ a gas	netic wave

The differentiating characteristics, C, provide information about A, usually by answering the following questions:

1. What are A's characteristics?
2. What is it composed of?
3. How does it work?
4. What does it do?
5. Where is it used/found?
6. When is it used?
7. What does it resemble?
8. Who uses/discovered it?

EXERCISE 2.2   Make complete definitions of the species (A) in Exercise 2.1 by answering one or more of the questions above.

Example: <u>A</u> carbon → <u>B</u> element → Where is it used/found?
 Answer: Carbon is an element that is found in all organic molecules.

EXERCISE 2.3   Using the words given below, and keeping them in the same order, add the minimum number of words necessary to make a formal definition.

Example: (carton/container/paper)
 Answer: A carton is a container that is made of paper.

1. (coal/substance/carbon)
2. (telescope/instrument/makes distant objects appear larger)
3. (electron/particle/mass of 9.107×10^{-28} grams)
4. (cube/geometric shape/equal sides)
5. (amoeba/animal/one cell)
6. (smog/atmospheric condition/caused in part by automobile exhaust)
7. (microphone/device/amplify sound)
8. (stainless steel/metal/not corrode)
9. (uranium/element/nuclear reactors)
10. (pipette/instrument/measures small amounts of liquid)

REMOVING THE GENERAL-CLASS WORD IN A DEFINITION

Sometimes, the B element (general-class word) is not necessary because it is repetitious or obvious:

> A Beckmann thermometer is a thermometer that measures small temperature differences.

In this definition, the general-class word *thermometer* is not helpful in defining the Beckmann thermometer. Therefore, the B section can be omitted:

> A Beckmann thermometer measures small temperature differences.

The general-class word in a definition may sometimes be omitted for a technical audience:

> Technical version:
> A nucleotide is formed from one molecule of a sugar, one molecule of phosphoric acid, and one molecule of a base containing an amino group.
> Less technical version:
> A nucleotide <u>is a chemical compound that</u> is formed from one molecule of a sugar, one molecule of phosphoric acid, and one molecule of a base containing an amino group.

EXERCISE 2.4   Draw a line through the general-class section (B) in the following definitions if it is not necessary. Explain why you removed it.

Example: An electrical switch is a device that opens and closes a circuit.
Answer: An electrical switch ~~is a device that~~ opens and closes a circuit.
Explanation: The general-class section is not necessary because the word *device* is too general and does not really help us to understand the meaning of *an electrical switch*.

1. A suspension bridge is a bridge that is supported by cables.
2. Irrigation is a process that supplies water to the soil in dry areas.
3. A stoma is a small pore that is located in the epidermis of a leaf.
4. A rheostat is a variable resistor that varies the current in a circuit.
5. A gas thermometer is a thermometer that uses gas as the working substance.
6. A red blood cell is a flattened, disc-shaped cell that circulates in vertebrate blood.
7. A beam is a long piece of wood or metal that is supported at both ends and often used in constructing buildings.

8. A flower is a plant's mechanism for attracting birds and insects for pollination.
9. A light-year is a unit of distance that is equal to 6×10^{12} miles.
10. Soap is a substance that suspends dirt particles.
11. Tungsten (W) is an element that is used for light-bulb filaments.
12. Titration is a process that determines the pH of a liquid.
13. $C = 2\pi r$ is a mathematical formula that determines the circumference of a circle.
14. Cilia are filaments that enable some one-celled animals to move around.
15. Gravity is a force that can deflect light.
16. A transformer is a device that alters the voltage of alternating current.

Section 3
SENTENCE COMBINING: DEFINING RELATIVE CLAUSES

In section 2, we saw that the differentiating characteristics (part C) of a definition usually begin with the word *that*. A definition is really two sentences that have been combined:

A thermometer is ⌐an instrument.⌐   (main sentence)
+   THAT
⌐An instrument⌐ measures temperature.   (clause sentence)

A thermometer is an instrument that measures temperature.

We use this combined form, which is called a defining relative clause, because we want to avoid repeating the words *an instrument*. In the clause sentence, the subject of the sentence (*an instrument*) is replaced by the relative pronoun *that*.

Defining relative clauses are not only used in definitions. They are used in any sentence in which we want to give necessary information about a noun. Look at this sentence:

Pictures <u>that showed the rings of Saturn</u> were presented.

If we remove the defining relative clause, *that showed the rings of Saturn*, from the sentence, we have

Pictures were presented.

This sentence alone tells us very little. We need the defining relative clause in order to give the noun *pictures* a meaning.

EXERCISE 3.1   Combine the following sentences to make a sentence with a defining relative clause. Make the first sentence the main sentence.

Example: Pictures were presented. Pictures showed the rings of Saturn.
Answer: Pictures that showed the rings of Saturn were presented.

1. The planet is Mercury. The planet is closest to the sun.
2. The movie was *Star Wars*. The movie won the most awards.
3. The element is berkelium. The element was discovered at U.C. Berkeley.
4. The device was the telephone. The device started modern electronic communication.
5. The organs are the kidneys. The organs are concerned with the elimination of nitrogenous wastes.

In a definition, it is possible to use the relative pronoun *that* for both things and human beings:

A circuit is a complete path <u>that</u> can be followed by an electric current.
A pediatrician is a doctor <u>that</u> attends to children.

However, in defining relative clause sentences that are not definitions, we usually use the relative pronoun *who* or *whom* for human beings, especially when there is a personal relationship:

The student <u>who</u> sits next to me is an engineer.
The man <u>whom</u> they met was studying earthquakes.

Who is the only relative pronoun that has a subject form (*who*) and an object form (*whom*):

The patient is undergoing chemotherapy.   (main sentence)
+   **WHO**
The patient spoke to us.   (clause sentence)

The patient who spoke to us is undergoing chemotherapy.

The patient is undergoing chemotherapy.   (main sentence)
+ ──────────────**WHOM**
Dr. Smith examined the patient.   (clause sentence)

The patient whom Dr. Smith examined is undergoing chemotherapy.

These two forms will be discussed after Exercise 3.2.

EXERCISE 3.2   Add the correct relative pronoun—*that, who,* or *whom*—to the following sentences.

Example: The man _____ they met was studying earthquakes.
Answer: The man whom they met was studying earthquakes.

1. A paleontologist is a person _____ studies fossils.
2. My sister _____ lives in New York is studying computer science.
3. The textbook _____ we use in this class costs twenty-five dollars.
4. Our office needs somebody _____ can type at least 100 words a minute.
5. The people _____ need help the most are those _____ have no food or shelter.
6. My cousin _____ studies at Harvard is investigating the great galaxy in Andromeda.
7. The software _____ we really need is a word-processing program.
8. The woman _____ he heard at the conference is a well-known particle physicist.
9. The men _____ first described the DNA double helix are Watson and Crick.
10. Men and women _____ attend to patients in a hospital are called nurses.

SUBJECT-FORM AND OBJECT-FORM DEFINING RELATIVE CLAUSES

Look at this example from Exercise 3.1:

> The planet is Mercury.   (main sentence)
> +   **THAT**
> The planet is closest to the sun.   (clause sentence)

(1) The planet that is closest to the sun is Mercury.

The words *the planet* in the clause sentence are the **subject** of the sentence. For this reason, sentence 1 contains a **subject**-form (or S-form) relative clause. If we use a different clause sentence,

> The planet is Mercury.   (main sentence)
> +┌──────────────────── **THAT**
> ▼The astronomers are observing the planet.   (clause sentence)

(2) The planet that the astronomers are observing is Mercury.

the repeated words *the planet* are now the **object** of the clause sentence. For this reason, sentence 2 contains an **object**-form (or O-form) defining relative clause.

EXERCISE 3.3   Are the defining relative clauses in the following sentences subject-form (S) or object-form (O)?

Example: S The planet that is closest to the sun is Mercury.

1. _____ One scientist who received the Nobel prize for chemistry was Dr. Paul Berg of Stanford.
2. _____ The nurse whom the doctor hired was very efficient.
3. _____ The food that you eat should be varied.
4. _____ Invertebrates are animals that have no backbone.
5. _____ The new pilot had to fly a plane that he had never flown before.
6. _____ The air pollutant that is most responsible for dirty air is carbon monoxide (CO).
7. _____ Test-tube babies are children that are conceived outside the mother's body.
8. _____ A new agent that scientists hope will cure cancer is interferon.

In many scientific sentences, the relative clause indicates a condition or circumstance, and the main sentence shows the result of the condition or circumstance (see Unit II, Section 2, "Main Clause versus Subordinate Clause"):

CONDITION OR CIRCUMSTANCE   RESULT

Butter that is left in the sun usually melts.

EXERCISE 3.4   In the following pairs of sentences, decide which is a result (R) or principle and which is a condition (C) or circumstance. Make the condition sentence into an S-form defining relative clause and combine the sentences.

               R   C
Example: Water becomes steam. Water is heated to 212° F.
 Answer: Water that is heated to 212° F becomes steam.

1. An illness can usually be cured by antibiotics. An illness is caused by bacteria.
2. An iron bar is exposed to moisture. An iron bar rusts.
3. A dirigible is a large commercial balloon. A large balloon is filled with helium (He).
4. The nerves do not respond to touch or pain. The nerves are in the brain.
5. Chlorine (Cl) is added to water. Water is used for drinking.
6. Metal is covered with paint or plastic. Metal is more resistant to corrosion.
7. Crystals generally have high melting points. Crystals are formed from ionic compounds.

8. A vitamin controls root growth. A vitamin is made in the leaves and sent down to the roots.
9. Some viruses alter the genes. The genes control the multiplying of cells.
10. A dam requires less concrete but more labor. A dam is used in rocky mountain areas.

EXERCISE 3.5   Combine the following pairs of sentences into a sentence containing an O-form defining relative clause. Make sentence (a) the main sentence and sentence (b) the clause sentence.

Example: (a) The element was radium.
(b) Marie Curie discovered radium.
Answer: The element that Marie Curie discovered was radium.

1. (a) The image is cast upside down on the retina of the eye.
   (b) We see the image.
2. (a) The total amount of hydroelectric power was approximately 5×10^9 kilowatt hours.
   (b) Britain generated the hydroelectric power in 1970.
3. (a) Smog formation is a process.
   (b) Scientists are investigating the process.
4. (a) One food-preserving process is called freeze-drying.
   (b) The food industry uses one food-preserving process on a large scale.
5. (a) The patient complained of lower-back pain.
   (b) The doctor examined the patient.
6. (a) The male insects were attracted to the females by their scent.
   (b) We studied the male insects.
7. (a) Most ultraviolet light is absorbed by the ozone (O_3) layer.
   (b) Ultraviolet light forms the ozone layer.
8. (a) The amino acids have an amino group ($-NH_2$) at one end and a carboxyl group ($-CO_2H$) at the other.
   (b) Every organism produces the amino acids.
9. (a) The computer has been "down" (not working) for two weeks.
   (b) The EECS department installed the computer.
10. (a) The town was rebuilt in two years.
    (b) The hurricane destroyed the town.

Section 4
VERBS: *BE, HAVE,* AND THE SIMPLE PRESENT TENSE

We have seen that the most common verb in a definition is *be. Be* is very important in scientific writing. It is the main verb in about 35 percent of all scientific statements.

EXERCISE 4.1   Add *is* or *are* to the incomplete sentences in the following paragraph.

GLACIERS

A glacier like a history of the weather. In some areas, such as the Antarctic, glaciers in fact the only source of information we have. The records that they leave called moraines. A moraine a mass of rock and soil that carried along by the glacier. The age of these moraines identified by the plant growth on them or by carbon 14 dating methods. One cycle of a glacier's advance and retreat 30,000 years or more. We know that today we in an interglacial period.

A typical English sentence has the following structure:

SUBJECT	VERB	OBJECT	ADVERBIAL

It is not possible to make a complete sentence in English without a verb. However, in scientific writing the subject and the object of the sentence usually carry more information than the verb. The verb is usually very simple. It is often *be* or *have* or a single verb in the simple present tense. Look at these examples:

1. The spiral motion of air above a low-pressure area <u>is</u> always in a counterclockwise direction.
2. The investigators <u>have</u> preliminary results demonstrating endorphin-like material in tetrahymena.
3. The reaction in rabbits to an albumin-solution injection from the frozen carcass of a woolly mammoth <u>shows</u> the ancestral relationship between these two mammals.

These three sentences all have very simple verb structures: *be, have,* and the simple present verb *show.*

EXERCISE 4.2   Add the correct form of *be* or *have.*

1. A small protein molecule _____ a molecular weight of about 10,000.
2. Light _____ a speed of 186,000 miles per second.
3. There _____ several kinds of vitamins.
4. Five hundred pounds _____ the maximum weight that this rope can support.
5. Laboratory equipment _____ usually made of Pyrex glass.
6. Good stereo systems _____ a frequency range of 30 to 18,000 cps (cycles per second).
7. Gold coins usually _____ a composition of 90 percent gold (Au) and 10 percent copper (Cu).

8. Genetics ————— the study of the variation in heredity.
9. Although similar to hemoglobin, chlorophyll ————— magnesium rather than iron as the central atom.
10. Fractional distillations ————— methods of separating portions of different volatility from a liquid.

EXERCISE 4.3 Complete the fragments (incomplete sentences) by inserting the appropriate form of *be* or *have*.

Example: Adult human beings thirty-two teeth.
 Answer: Adult human beings have thirty-two teeth.

1. The boiling point of water 212° F.
2. Carbon monoxide and sulfur oxides the most common air pollutants.
3. Invertebrates no backbones.
4. Interferon an agent that might cure cancer.
5. Benzene six carbon atoms and six hydrogen atoms.
6. Our solar system nine planets.
7. Scientific textbooks quite expensive.
8. Paleontology the study of fossils and other forms of life.
9. Iron and potassium both ten isotopes.
10. An evergreen tree leaves all year round.

THE AGREEMENT BETWEEN SUBJECTS AND VERBS

The agreement between subjects and verbs is shown by *person* in grammar.

	Singular	Plural
First person:	I like tea.	We like tea.
Second person:	You like tea.	You like tea.
Third person:	He/She/It likes tea.	They like tea.

The third-person form of the simple present tense is very common in scientific writing because it is used to describe facts, processes, and observations. One easy way to remember the important *-s* at the end of this simple-present verb form is shown below. There must be one *-s*, but not two. If the *-s* is with the verb (example 1), then it is not with the subject. If the *-s* is with the subject (example 2), then it is not with the verb. Remember, the *-s* must be with either the subject or the verb—unless, of course, the subject has an irregular plural form, such as *men* or *people*.

(1) The tree grow**s**.
(2) The tree**s** grow.

IDENTIFYING THE "BARE" SUBJECT

Because the subject of a scientific statement carries so much information, the "bare" or simple subject is usually surrounded with articles and adjectives and other phrases. The bare or simple subject is very important because it is this word alone that controls the verb.

Satellite pictures of the sea surface show the sea floor in detail.

Radioactive radon cycled through a burning cigarette is a likely carcinogen.

EXERCISE 4.4   Underline the subject word or phrase in the following sentences. Put a second line under the single word in the subject that shows you whether the verb should be singular or plural. Write the correct form of the verb in the blank.

Example: The first element in the periodic table (be) ———— hydrogen.
  Answer: The first element in the periodic table (be) __is__ hydrogen.

    1. Water (boil) ———————————— at 100° C.
    2. An astronomer (observe) ———————— the universe through optical and radio telescopes.
    3. Red blood cells (circulate) ———————— in the circulatory system.
    4. The corrosion of silver (produce) ———————— silver sulphide.
    5. A rheostat (vary) ———————— the current in a circuit.
    6. The stomata in a leaf (open) ———————— and (close) ———————— to allow the exchange of oxygen and carbon dioxide.
    7. Children usually (resemble) ———————— their parents.
    8. A well-equipped laboratory (supply) ———————— water, air, and steam.
    9. Many foreign students (major in) ———————— technological fields.
    10. The nervous sytem (be) ———————— a mechanism by which an animal (perceive) ———————— and (respond to) ———————— its environment.

EXERCISE 4.5   Underline the word that controls the subject. Place the verb phrase in parentheses in the correct position in the sentence.

Example: The standard volume of a mole of gas at standard temperature and pressure 22.4 dm³. (be)

Answer: The standard <u>volume</u> of a mole of gas at standard temperature and pressure is 22.4 dm³.

1. Many cells in the brain area called the hypothalamus input from the taste sensors. (receive)
2. A frog a halfway stage between water and land creatures. (represent).
3. A bimetallic thermometer the different coefficients of linear expansion of two strips of metal riveted together. (depend on)
4. A person who is hit hard on the head consciousness. (often lose)
5. The effect in which a free electron can pass through a potential energy barrier even though it does not have energy to get past that barrier the tunnel effect. (be called)
6. The capillary tube in a clinical thermometer a narrow constriction near the bulb to prevent the mercury from moving after the patient's temperature is taken. (always have)
7. The class Reptilia millions of years ago from the class Amphibia. (evolve)
8. A variation, $d\Phi$, in the time, dt, of the flux linked with an electrical circuit an electromagnetic force in the circuit whose value is

$$E = -\frac{d\Phi}{dt} \quad . \quad \text{(have)}$$

9. Partly decomposed and unaltered bedrock beneath the zones of accumulation and leaching in a typical soil profile. (lie)
10. Growth form, natality, mortality, and population density the size of a population. (determine)

EXERCISE 4.6   In the three example sentences on page 18 of this section, which word controls the number (singular or plural) of the verb? There is only one in each case.

Section 5
WRITING AIDS: CAPITALIZING TITLES

Every formal paper requires a title, just as every graph and figure requires a label. In titles, every word has a capital letter except for four kinds of words.

EXERCISE 5.1   Analyze the following titles for the correct use of capitalization. What kinds of words are not capitalized? Are there any exceptions? Can you make a general rule for capitalization in titles?

1. A Pictorial Guide to Fossils
2. Building a Computer

3. The Comet Is Coming!
4. A Grammar of Contemporary English
5. Reading Between the Lines
6. Mathematical and Logical Paradoxes
7. The World We Live In

In titles, every word is capitalized except articles (*a, an, the*), conjunctions (*and, but, or*), the verb *be* (*is, are, was, were*), and short prepositions (*in, of, by, at,* etc.). Of course, if any of these types of words come at the beginning of the title, they are also capitalized. Short prepositions and forms of *be* at the end of a title are also capitalized. Long prepositions such as *beyond, opposite, along,* and *through* are capitalized too.

EXERCISE 5.2   Capitalize the following titles correctly.

1. the chemicals you and i depend on
2. are we alone? the possibility of extraterrestrial civilizations
3. the origin of consciousness in the breakdown of the bicameral mind
4. we don't know who we are
5. the birds in the deserts of north america
6. the principles geometry is based on
7. the known universe and beyond
8. the algal bowl: lakes and man
9. the use of a portable shear-stress machine in the determination of shear-strength parameters
10. an investigation of the effect of radioactive labeling of DNA on excision repair in UV-irradiated human fibroblasts

PART II
Writing an Amplified Definition

Section 6
PREWRITING ACTIVITY: DESCRIBING A DIAGRAM[1]

EXERCISE 6.1 Choose *one* of the diagrams that follow on pages 23–26. You will have fifty minutes to write a composition of approximately 150 words describing the diagram you have selected. Be sure to check your writing before you give your paper to your instructor.

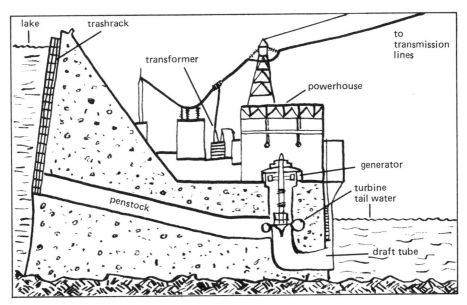

A Hydroelectric Dam

[1]*Note to teacher:* If this exercise is used as a diagnostic, it should be done at the beginning of the course and students should work by themselves. If it is to be an introduction to the general process of writing, students should be encouraged to discuss their individually selected diagrams with each other in pairs or groups and/or with you in a brainstorming session so that they can enhance their understanding of vocabulary, structure, process, etc., before they begin to write.

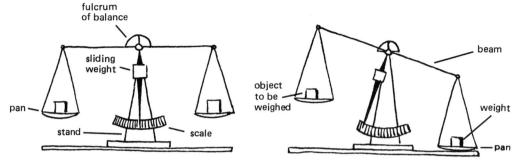

An Equal-Armed Balance Scale

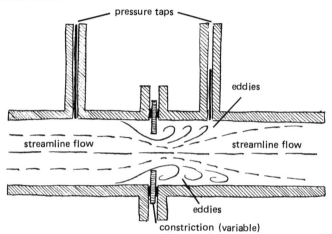

The Flow of Liquid Through a Pipe

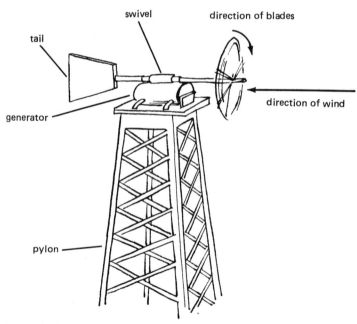

A Wind Generator

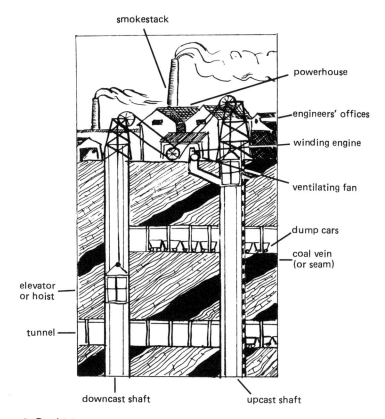

smokestack

powerhouse

engineers' offices

winding engine

ventilating fan

dump cars

coal vein
(or seam)

elevator
or hoist

tunnel

downcast shaft

upcast shaft

A Coal Mine

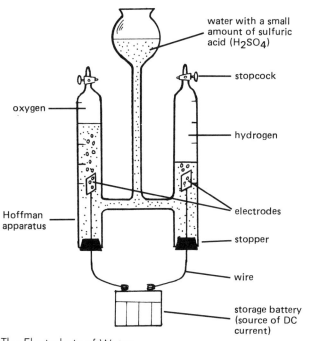

water with a small
amount of sulfuric
acid (H_2SO_4)

stopcock

oxygen

hydrogen

electrodes

Hoffman
apparatus

stopper

wire

storage battery
(source of DC
current)

The Electrolysis of Water

Writing an Amplified Definition   **25**

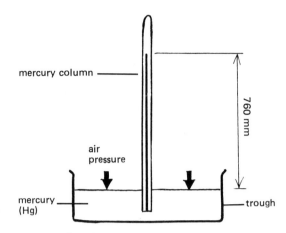

A Torricellian Barometer

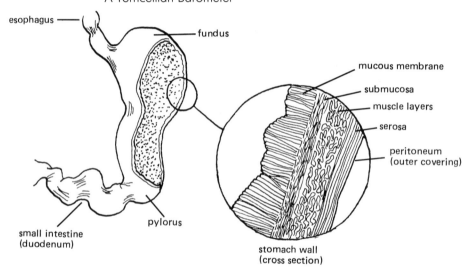

The Stomach

Section 7
STRUCTURE

An amplified definition is used in scientific writing when a simple, formal definition is not sufficient. Look at the following examples:[2]

1. A vaccine is a sterile liquid medium that contains an avirulent strain of a specific pathogen.
2. An n-type semiconductor is a type of semiconductor in which most of the current is carried by electrons rather than holes.

18
19

[2]Definitions from *Longman Dictionary of Scientific Usage* (London: Longman Group Ltd., 1979), pp. 281, 369, and 535.

3. Thixotropy is the property of a liquid by which it has a lower viscosity at a higher rate of flow.

Without specific experience in medicine, electronic engineering, or fluid mechanics, the reader of such definitions will probably need more information in order to understand exactly what these words mean. There are many ways to amplify or expand a definition. Here are the ten most common techniques:

1. Further definition.
   In example 1 above, we would understand the word *vaccine* better if we understood the words *avirulent* and *pathogen*. In other words, this definition would be amplified if these words were defined.
   In example 2, we need to know the specific meaning of the word *holes* in this definition.
   In example 3, we need to understand the word *viscosity*.
2. Concrete examples and instances.
   Example 1 would be clearer if we knew that vaccines are used to protect living organisms from certain diseases, such as tuberculosis and rabies.
   Example 3 would be clearer if we knew that thixotropy is a property of certain gels and paints.
3. Description of parts or components.
   In example 2, it might help to know that a semiconductor is composed of elements or compounds such as germanium, selenium, silicon, and lead telluride. This knowledge would be especially useful if we were concerned with the manufacture of semiconductors.
4. Basic operating principle.
   In example 2, we need to know how semiconductors work, how current is "carried," and so on.
5. Purpose or method of use.
   If the word being defined has a use, we need to know what it is used for and how it is used.
   In example 1, vaccines are commonly injected.
   In example 2, n-type semiconductors are used in transistors and other electronic components.
6. Cause and effect.
   In example 1, what is the result (effect) of using a vaccine?
   In example 3, we need to know that movement causes a lower viscosity in thixotropic liquids.
7. Word derivation.
   In example 2, *n-type* would be clearer if we knew that *n* means *negative*.
   In example 3, we would understand *thixotropy* better if we knew that the word is derived from two Greek roots: *thixo* ("touch") and *trope* ("turn toward or away from"). If we stir or agitate (touch) a thixotropic liquid, it becomes (turns) more liquid.

8. Location and time.
   In example 1, when and where would a vaccine be used?
   In example 2, where would you expect to find an n-type semiconductor?
9. Negative statement.
   In example 3, thixotropy is not a property of most liquids.
10. Comparison and contrast.
    In example 1, a vaccine can be compared and contrasted with an antiserum.
    In example 2, an n-type semiconductor can be compared with a p-type semiconductor.
    In example 3, thixotropy can be compared and contrasted with dilatancy.

EXERCISE 7.1   With another student, choose three techniques from the above list that you could use to amplify a definition of the words below.

Example: a mercury thermometer

1. Technique 4: Basic operating principle:
   Mercury expands when heated and rises in the capillary a distance proportional to the temperature increase.
2. Technique 5: Method of use.
   The bulb of the thermometer must be surrounded by the substance being measured.
3. Technique 10: Comparison and contrast.
   Mercury thermometers can be contrasted with other types (Beckmann, ethanol, bimetallic, gas, etc.).

a. a photoelectric cell
b. the solar system
c. a red blood cell
d. an isotope
e. photosynthesis

f. a gamete
g. corrosion
h. a venturi meter
i. radar
j. a vitamin

k. a bulldozer
l. a Bunsen burner
m. a pollutant
n. an electromagnetic wave

Section 8
MODELS

MODEL 1: AN ANEROID BAROMETER

An aneroid barometer (Figure 1.2) is an instrument that depends on the changing volume of a container to indicate atmospheric pressure. It consists of an airtight box of thin flexible metal from which the air has been partially evacuated. One side of the evacuated box is attached to a spring. When atmospheric pressure increases, the box tends to collapse. When atmospheric pressure decreases, the sides of the box spring outward. This slight movement

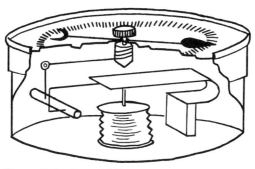

Figure 1.2 Aneroid Barometer

is magnified by a series of levers connected to an indicator needle, which shows the atmospheric pressure.

A variation of the aneroid barometer called the Bourdon gauge was invented by Eugene Bourdon, a French engineer. A flattened tube of metal is evacuated and bent into a circle. The circle tends to close up with greater pressure and open out with lesser pressure. This movement is transmitted to a dial as in the aneroid instrument. The Bourdon gauge is most suitable for measuring high pressure (for example, 2000 atmospheres).

MODEL 2: LATEX

Latex is a milky, clear, or sometimes colored liquid that exudes from the cut surface of certain flowering plants and that coagulates rapidly on exposure to air. It is thought to be concerned in the protection and healing of superficial wounds to the plant and also in its nutrition. Latex contains many different substances, including sugars, proteins, mineral salts, alkaloids, and caoutchouc, a substance that is found in the latex from rubber trees (Figure 1.3).

Figure 1.3 Gathering Latex

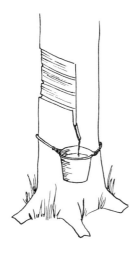

When latex is mixed with carbon and vulcanized (heated with sulfur), it forms commercial rubber, the most common latex product. Commercial rubber is used most importantly in the production of automobile and truck tires. The search for a synthetic rubber substitute was the forerunner of the enormous present-day plastics industry.

MODEL 3: THE CANINE TEETH

The canines are the teeth in the human dentition that are located between the incisors and the bicuspids, one in each quadrant (see Figure 1.4). Their function is to guide the teeth in chewing and to maintain proper facial contours by holding up the corners of the lips. The canines are the strongest teeth in the mouth and are therefore often used as anchors for partial dentures or bridges. They also have the longest roots of all the teeth and are the most impervious to decay. For this reason, they are the last teeth lost to gum disease.

The word *canine* comes from the Latin word for *dog*. Canines in animals cut and tear food and also serve as a defense mechanism. In some animals, the canines function as a whetstone, continuously sharpening the teeth as they mesh together in chewing. Saber-toothed tigers, now extinct, had the most developed canines in the animal kingdom.

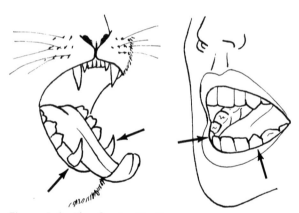

Figure 1.4 The Canine Teeth

Section 9
ANALYSIS

A typical amplified definition has the following structure:

<u>Title</u>
1. a formal definition
2. three or more amplification techniques
3. a description of special uses, more complex types, etc.

An amplified definition of a mechanical device, for example, might have the following structure:

<u>Title</u>

1. a formal definition
2. a description of functional parts ⎫
3. the basic operating principle   ⎬ amplification techniques
4. the method of use   ⎭
5. special uses, more complex types, etc.

This structure may be used in defining a mechanical device or anything that has a function or use and consists of separate components. All amplified definitions begin with a formal definition, but the means of amplification will, of course, depend on the subject being defined. The discussion of special uses, more complex types, etc., is a good way to end the amplified definition, especially if it is a self-contained composition and not part of a larger report.

EXERCISE 9.1   With another student in your field, analyze Model 1 in Section 8. Determine the function of each sentence in the model (title, definition, amplification techniques, special uses), identify the amplification techniques used (for example, method of use, cause and effect, word derivation), and label each sentence in the margin. Discuss your results with your partner. Do you agree?

EXERCISE 9.2   Analyze Model 2 and Model 3 as you did with Model 1 in Exercise 9.1. Compare your analyses with that of Model 1. Are they different? If so, how are they different?

Section 10
CHOOSING A TOPIC

Your assignment is to write an amplified definition in your own field. First you must choose a topic.[3] Many examples have been given in this unit that are good topics for an amplified definition (a venturi meter, photosynthesis, etc.). Here are some additional topics related to certain majors.

Biology/Plant Pathology	Civil/Mechanical Engineering
Annual Rings	Concrete
Amino Acids	A Foundation
The Golgi Apparatus	The Lever
The Paramecium	A Theodolite

[3]*Note to teacher:* Give students as much help in deciding on a topic as you find necessary. Have them use resources such as the library, and possibly radio and television programs. You might want to encourage students in similar fields to make rough outlines, exchange them with each other, and then discuss their strengths and weaknesses, using you as a resource person. This would help to focus the reason for a visit to the library or the use of some other resource.

Electrical Engineering/ Computer Science	Chemistry/Chemical Engineering
The Automotive Battery	A Heat Exchanger
A Diode	A Glass Electrode
The Multimeter	A Distillation Column
A Transistor	A Titrimiter

Medicine/Physiology	Physics/Astronomy
The Electrocardiagram	The Laser
A Hemodializer	A Pulsar
Polysaccharides	The Periodic Table
The Killer T Cell	A Spectrum

EXERCISE 10.1   Write a one- or two-page amplified definition using the structure shown in Section 9. Remember that your first sentence must be a formal definition and that your last section should concern special uses or more complex types. (Some of you will be able to write a composition from your own knowledge. Others will have to use the library and books or journals in your field. Ask your teacher if you need help in finding reference material.)

UNIT II
The Description
of a Mechanism

PART I
Grammar

Section 1
ARTICLES: THE DEFINITE ARTICLE

FIRST VERSUS SECOND MENTION

Before we analyze the use of the definite article, try the following exercise.

EXERCISE 1.1 Fill the blanks with a(n), ∅, or *the*. See Figure 2.1 on page 34.

MAKING _____ DISTILLED WATER

_____ distilling flask is attached to _____ condenser, as shown in the diagram. _____ condenser is connected to _____ source of cold water and _____ beaker is placed below it. _____ flask is partially filled with _____ water. _____ thermometer is inserted through _____ stopper (_____ stopper seals _____ flask). _____ flask is heated with _____ Bunsen burner. When _____ water boils, it becomes _____ steam. _____ steam rises in _____ flask and passes through _____ condenser, where it is cooled. _____ condensed steam collected in _____ beaker is distilled water.

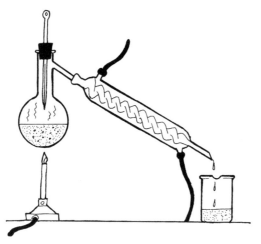

Figure 2.1 Distillation Apparatus

Analysis: Draw a (circle) around each noun that is preceded by *the*. Can you find this noun in an earlier sentence in the paragraph? If you can, draw a box around the earlier noun.

The first time a noun is mentioned in a paragraph is called first mention. We usually use the indefinite article (as discussed in Unit I, Section 1) for first mention, according to the following rule:

a(n) + a singular countable noun (e.g., *a book*)
  Ø + a plural countable noun (e.g., *books*)
  Ø + an uncountable noun (e.g., *air*)

The second time a noun **with the same reference** is mentioned in a paragraph is called second mention. Second mention includes any subsequent use of a noun with the same reference (that is, a noun that refers to the same noun). We use the definite article *the* for second mention, according to the following rule:

the + a singular countable noun (e.g., *the book*)
the + a plural countable noun (e.g., *the books*)
the + an uncountable noun (e.g., *the air*)

First mention and second mention show different views of a noun in English. First mention is similar to taking a picture of an object that is far away. Second mention is similar to using a zoom lens on your camera to bring the object close to you. In other words, we first identify a noun with an indefinite article (*a(n)* or Ø), and after that we keep it identified with the definite article

Figure 2.2 First Mention versus Second Mention

(the). Therefore, if you change the article back to first mention, the reader or listener will understand that you mean a **different** example of the same noun (in other words a different reference). Look at this example:

> (1) An old man is walking with a small boy. (2) The man is tired, but the boy is hungry and needs some food. (3) The boy has not eaten for several days. (4) He looks thin and unhealthy. (5) Suddenly, the old man stops. (6) A man calls out, "Would you like some bread?" (7) The man smiles.

Notice that we identify *an old man* (first mention) in sentence 1 and repeat *the man* in sentence 2 and sentence 5, understanding that it is the same man each time. When we see *a man* in sentence 6, we immediately understand that a different man has appeared. The last sentence is not a good sentence because we do not know **which** man is meant, the old man or the bread man? How would you correct the last sentence to make it clear?

Sometimes a second mention appears not as the same noun but in the form of a synonym or a defining phrase:

> Dr. Lee bought a new Mercedes. The car is silver with a black interior.
>
> The nurse ordered a sleeping pill for the insomniac patient. The medication was very effective.

EXERCISE 1.2 Fill the blanks with *a(n)*, *the*, or *∅*.

SOAP

_____ simple experiment demonstrates how _____ alkali acts on _____ grease or _____ oil. Put _____ spoonful of _____ washing soda with _____ water in _____ greasy frying pan. Boil _____ mixture. In _____ short time, _____ washing soda and _____ grease become

_____ particles. _____ particles unite to form _____ new substance that we call soap. _____ soap can be washed out, leaving _____ frying pan clean.

SECOND MENTION WITHOUT FIRST MENTION

Second mention can occur without first mention if (1) a ranking adjective occurs before a noun and (2) if shared knowledge removes the need for first mention.

Ranking Adjectives

The rule for first and second mention is not required in English if a noun is preceded by a ranking adjective. There are three types of ranking adjectives: superlative, sequential, and unique. They all make the noun definite and thus require *the*.

Superlative adjectives. The superlative adjectives indicate the strongest form of comparison (see Unit VI, Section 3, "Comparisons").

This is the best solution.
Dioxin is one of the most notorious pollutants.
Hydrogen has the lowest atomic weight, uranium the highest.

Sequential adjectives. The sequential adjectives indicate sequence in time or space. They include *the first, the second, the next, the following, the last,* etc.:

The first man on the moon was Neil Armstrong.
The second phase of an internal-combustion engine is called compression.
The following are prime numbers: 1, 3, 5, and 7. What is the next?
The last glaciation ended 10,000 to 15,000 years ago.
Nuclear weapons were used in the last month of World War II.
The next month, the war was over.

EXCEPTION: When referring to points in time from the present, we use *next* or *last* with no article:

The cyclotron will be completed next month (from now). Last month, it was in the final stages.

Unique adjectives. The unique adjectives indicate that only one such noun is possible. They include *the same, the only, the one, the chief, the principal, the main, the whole, the entire, the complete,* etc.:

Researchers are investigating <u>the principal</u> causes of cancer.

Natural oil is <u>the one</u> product on which the industrialized world depends.

<u>The only</u> specification was that the paint should dry in twenty-eight minutes.

EXCEPTIONS: <u>an</u> only child = a child with no brothers or sisters

<u>a</u> chief

<u>a</u> principal   cause = there are several major causes

<u>a</u> main

EXERCISE 1.3   Write the correct form of the ranking adjective in parentheses in the blank.

1. The Soviet Union is (large) _____ country in the world.
2. Some arctic rocks have (same) _____ composition as rocks from Mars.
3. (first) _____ person to discover the application of penicillin was Alexander Fleming.
4. The Pacific "ring of fire" is the location of (extensive) _____ _____ earthquake activity in the world.
5. Marsupials (e.g., kangaroos) are (only) _____ animals to carry their young in a pouch.
6. Pollution is (principal) _____ factor in smog formation.
7. (light) _____ metal used for construction is aluminum.
8. (only) _____ child faces different problems than a child with brothers and sisters.
9. A new oil policy is going to be implemented (next) _____ year.
10. An embryo is (final) _____ stage in the development of a zygote.

Shared Knowledge

A second condition that does not require the rule of first and second mention is shared knowledge. The shared-knowledge rule states that if a noun is identified because we know that there is only one of it or because we immediately know which one is meant, we always use the article *the*. Look at this example:

<u>The sun</u> warms <u>the ocean</u>.

We know which sun and which ocean (the whole ocean, not one of the individual ones) are meant. If we said,

> *A sun warms the ocean.

we would think that we were on a different planet or in a different solar system with two or more suns. On this planet, everyone in the world shares the knowledge that there is only one sun.

There are three different levels of shared knowledge: (1) world, (2) cultural, and (3) regional/local.

World shared knowledge. World shared knowledge applies to nouns that everyone in the world knows. These include *the sun, the moon, the earth, the ground, the sky, the ocean, the universe, the cosmos,* etc.

EXCEPTIONS: We do say not *the life* but rather *life*, not *the nature* but *nature*, unless we intend to limit these nouns (see "Articles with Postmodified Noun Phrases," Unit 3, Section 1):

> the life of a poor person
> the nature of man

EXERCISE 1.4   Answer the following questions with complete sentences.

1. What is the distance between the sun and the earth?
2. Why is the sky blue?
3. What is the big-bang theory of the formation of the universe?
4. If an object falls from a height of one kilometer, what is its velocity when it hits the ground?
5. What is the average concentration of salt in the ocean?
6. What is the diameter of the moon?
7. In what other place in the galaxy has nature provided the conditions required for life?

Cultural shared knowledge. Cultural shared knowledge applies to concepts that everyone in a certain culture knows. In the Westernized cultures, these include *the telephone, the radio, the capital, the movies, the theater, the (news)paper, the time* (that is, *the present hour*), etc.

EXCEPTIONS: Many of these "cultural concept" nouns can also be single items for which we do not use cultural shared knowledge. For example, we say:

> There's a movie on television (concept).
> not *There's a movie on the television (object).

(We can say "There's a clock on the television," but this is local, not cultural, shared knowledge. See also Unit VI, Section 1, "Idiomatic Structures With ∅.") We also say:

> I had a nice time (an experience).
>
> Time is relative (abstract idea).
>
> I don't have time (free time).

EXERCISE 1.5   Write the correct form of the words in parentheses.

1. Answer (telephone) ————————————— , please.
2. They had (push-button telephone) ————————————— installed.
3. Have (good time) ————————————— at the picnic tomorrow!
4. Do you have (time) ————————————— , please?
5. If you can't afford a taxi, take (bus) ————————————— .
6. A car hit (bus) ————————————— in front of my house last week.
7. They're building (new theater) ————————————— next to (opera house) ————————————— .
8. We saw a new play at (theater) ————————————— last night.
9. There's a good science program on (TV) ————————————— tonight.
10. Put those flowers on (TV) ————————————— , please.

Regional/local shared knowledge.   Regional/local shared knowledge applies to nouns that everyone in a certain region or area knows. These might include *the university, the river, the city, the library, the park, the street,* and *the weather,* if such things exist in a certain region and if there is only one of each. Regional/local shared knowledge also applies to nouns in a room or a house. These might include *the kitchen, the bathroom, the garage,* and *the living room* in an apartment or a house; *the door, the window, the curtains, the floor, the light switch,* and other objects in plain sight in a room.[1]

EXERCISE 1.6   The following directions were written down in a telephone conversation. Reconstruct the directions, using complete sentences as the speaker on the telephone originally gave them.

---

[1]This "plain sight" aspect is often used in literature. The author creates a sense of "being there" by using second-mention *the* in a normally first-mention situation. For example, a story might begin like this: The storm has passed and the house is quiet now. The beach is bathed in orange as the late afternoon sun pierces the edges of the rain clouds.

1. Take Franklin Freeway to bridge
2. Cross bridge
3. Left at Court House
4. Right at university
5. Left at gas station
6. Around park
7. Right at library
8. Stop at second green house on right
9. Open door (it's unlocked)
10. Turn on light
11. Go through living room
12. Take stairs to back garden
13. John working in garage

In the instructions for the use of a device, one can *imagine* that an object is in plain sight. For example:

Remove <u>the drill</u> from <u>the box</u>. Plug in <u>the black power cord</u>.
Press <u>the red switch</u> to operate the drill.

EXERCISE 1.7   Fill the blanks with *a(n)*, *Ø*, or *the*.

FILM LOADING AND WINDING[2]

To avoid unnecessary delays when loading _____ film, set _____ shutter mode dial to "100X." If you must load _____ film with _____ dial set to "AUTO," remove _____ lens cap and aim _____ camera towards _____ bright light source. To open _____ camera back, lift _____ film rewind knob up sharply. Insert _____ film cartridge in _____ film chamber and lock _____ cartridge in place by returning _____ film rewind knob to its original position. Make sure _____ film is engaged properly on _____ spool by inserting _____ leader of _____ film at least the width of one perforation.

REVIEW OF THE ARTICLE

Indefinite

Use *a(n)* or *Ø*:

1. for first mention
2. for general characteristics
3. in definitions

34   [2]From *The Pentax MG Handbook*. Used with permission.

Definite

Use *the:*

1. for second mention
2. for nouns with ranking adjectives
3. for shared knowledge

EXERCISE 1.8   Add *a(n)*, *0*, or *the* where necessary.

HIGH SCHOOL CHEMISTRY EXPERIMENT

First time we saw dramatic chemical reaction was in our high school. Teacher instructed class to put 20 ml. of silver nitrate into beaker. We then took piece of copper wire and bent it into coil. Coil was suspended in beaker. Teacher told us to put beaker and copper coil into dark closet to keep it away from sun. We left experiment alone overnight. Next day, we opened closets and removed beakers. All of us had same reaction: we were all surprised. Clear silver nitrate solution had become turquoise blue like ocean. Silver nitrate had become copper nitrate. Coil had become very thin. And at bottom of beaker there were strands of precipitated silver. Teacher explained that copper atoms in coil had replaced silver atoms in solution, and silver atoms had been deposited as precipitated silver.

Section 2
SENTENCE COMBINING: DEFINING RELATIVE CLAUSES

MAIN CLAUSE VERSUS SUBORDINATE CLAUSE

In the relative clauses we studied in Unit 1, we identified two components: (1) the main sentence and (2) the clause sentence. We combined these two sentences to make a relative-clause sentence.

In this unit, we will call the main sentence the **main clause** and we will call the clause sentence the **subordinate clause**. How do we know which of the two precombined sentences should be the main clause and which should be the subordinate clause? Since the word *subordinate* means *less important,* we usually make the "more important" sentence the main clause and the "less important" sentence the subordinate clause. In other words, the sentence that is related less directly to the topic of the paragraph is usually subordinated.

In sentences that are concerned with causes and effects, this decision is easier to make. The effect or the result of a process must be the main clause, and the cause or conditions must be the subordinate clause. Look at this sentence:

Paint contains lead (Pb).   (condition/cause)
+ Paint is poisonous.   (result/effect)

Paint that contains lead is poisonous.

If we made the condition/cause sentence the main clause, we would get this sentence:

*Paint that is poisonous contains lead.

This sentence is not logical because it states that all poisonous paint contains lead (Pb), and we know that there are other poisons besides lead (for example, organic solvents, arsenic, and cyanide).

EXERCISE 2.1   Correct the following sentences if necessary by exchanging the position of the clauses.

Example: A boat that will not float has a hole in it.
 Answer: A boat that has a hole in it will not float.

1. The water that is usually taken from a river or a lake is used to cool a nuclear plant.
2. Most infections that are caused by bacteria can be cured with antibiotics.
3. Metal pipes that become severely corroded transport salt water.
4. Some fish that have no eyes live in the deepest parts of the ocean.
5. Light-bulb filaments that are made of tungsten last longer than other types.
6. Glass that can be blown into various shapes is melted.
7. Doctors who specialize usually earn higher salaries.
8. Plants that have the ability to store water survive well in the desert.
9. Many workers who develop cancer inhale asbestos dust.
10. Infrared waves that are blocked from reaching the earth are absorbed by water vapor in the atmosphere.

RELATIVE CLAUSES WITH PREPOSITIONS

If an object-form (O-form) relative clause (see Unit I, Section 3) contains a verb with a preposition, two constructions are possible:

(1) The substance that rubber is made from is latex.

Sentence 1 is common in spoken and written English but is considered informal. To make the sentence more formal and therefore more appropriate for scientific writing, we move the preposition in front of the relative pronoun:

(2) The substance <u>from</u> which rubber is made is latex.

This preposition movement is possible only with the relative pronouns *which* or *whom*. If the relative pronoun *that* has been used, it must be changed to *which* before the preposition can be moved.

The substance that rubber is made from is latex.
The substance from which rubber is made is latex.
Incorrect: *The substance from that rubber is made is latex.

EXERCISE 2.2   Combine each of the following pairs of sentences into one sentence that contains a formal relative clause. Make (a) the main clause and (b) the subordinate clause.

Example: (a) The planet is the earth.
(b) We live on the planet.
Answer: The planet on which we live is the earth.

1. (a) The temperature is called the melting point.
   (b) A substance changes from a solid to a liquid at the temperature.
2. (a) The subject is the treatment of arthritis.
   (b) This report is concerned with the subject.
3. (a) The man is the project engineer.
   (b) You should write to the man.
4. (a) The process is photosynthesis.
   (b) Plants depend on the process for usable energy.
5. (a) The organs are the lungs.
   (b) Oxygen is exchanged with carbon dioxide in the organs.
6. (a) Water cooling is the method.
   (b) The temperature of an internal-combustion engine is controlled by means of the method.
7. (a) The sun is the star.
   (b) The nine planets of the solar system orbit around the star.
8. (a) Epithelial tissue is the layer of skin.
   (b) Beneath the layer of skin lies the connective tissue.
9. (a) Electroplating is the process.
   (b) A thin layer of metal is bonded to another metal by the process.
10. (a) The aorta is the principal artery.
    (b) Blood flows to the brain through the artery.

EXERCISE 2.3   Make the following sentences formal by shifting the preposition.

Example: Latex is the substance that rubber is made from.
Answer: Latex is the substance from which rubber is made.

1. The doctor examined the lump that the patient was concerned about.

2. The article that the feasibility study referred to was no longer available.
3. The electrons that the nucleus is surrounded by have a very low mass.
4. The man that the group disagreed with presented some new evidence.
5. The barrier that the salmon jumped over was immediately heightened.
6. The method that the engineer disapproved of was quietly abandoned.
7. The muscle that the tumor was removed from was badly damaged by the operation.
8. There are numerous ways that solar energy can be used (in).
9. Darwin is a scientist that zoologists have great respect for.
10. The data that the researchers had relied upon turned out to be inaccurate.

RELATIVE CLAUSES WITH *WHOSE, WHERE* OR *WHEN*

The relative pronoun *whose* and the relative adverbs *where* and *when* are used less frequently than the other relative pronouns.

Whose

Whose is the possessive relative pronoun. Unlike the other relative pronouns, it must always be linked to a noun.

[The patient] is in room 72.

+   **WHOSE KIDNEY**   (S-form)

[The patient's kidney] was removed.

The patient whose kidney was removed is in room 27.

[The protein] is DNA.

+   **WHOSE STRUCTURE**   (O-form)

Watson and Crick discovered [the protein's structure.]

The protein whose structure Watson and Crick discovered is DNA.

Notice that the word *whose* is used with both people and things.

Where

Where is the relative adverb that indicates location. Unlike the relative pronouns *whom*, *which* and *whose*, relative adverbs cannot be preceded by a preposition, because relative adverbs **replace** prepositions. Furthermore, relative adverbs can be used only in object-form (O-form) relative clauses.

A sound studio is [a place.]

+   ——————————————————WHERE
    Ambient sound is controlled [in a place.]

A sound studio is a place where ambient sound is controlled.

Notice that the preposition *in* is also deleted when *where* is used.

Even though *where* indicates place, it is also possible to use *where* to indicate relative position. In this case, *where* is a reduced form of *wherein* (or the archaic word *whereat*), which is not limited to place but is considered extremely formal.

Note the point where the solid melts.
   (i.e., The solid melts *at this point*.)
This is a case where(in) only the parents can decide.
   (i.e., Only the parents can decide *in this case*.)

Where is also acceptable in defining the members of an equation or formula (see also Unit III, Section 5, "Referring to Equations"):

Einstein formulated the famous equation $E = mc^2$, where $E =$ energy, $m =$ mass, and $c =$ the speed of light.

When

When is the relative adverb that indicates time.

The Renaissance was [the period.]

+   ——————————————————— WHEN
    The sciences were revived [during the period.]

The Renaissance was the period when the sciences were revived.

Notice that the preposition *during* (or any other preposition of time) is deleted when *when* is used. This structure is not very common in scientific writing because it always requires a descriptive time phrase (e.g., *the time when, the point in history when, the exact moment when*). Because of the concise nature of scientific writing, we usually prefer to use an embedded question beginning with *when* (see "Embedded Questions," later in this section).

EXERCISE 2.4   Fill the blanks with *whose, where,* or *when.*

   1. Persons _____ blood pressure is high should not eat too much food containing sodium.
   2. The big-bang theory describes the moment nine billion years ago _____ the universe began.

3. The root is the section of a plant _____ water and minerals are absorbed.
4. Power, the rate of doing work, is described by the equation $P = w/t$, _____ $w = $ work done and $t = $ time taken.
5. The earth is a planet _____ core is a mixture of iron and lighter elements, such as sulfur and oxygen.
6. The period _____ some mammals sleep during the winter is known as hibernation.
7. Benzene is a molecule _____ structure is in the form of a ring.
8. The plane had reached a state _____ it could neither fly nor land safely.
9. A synapse is a point _____ the axon of a nerve cell comes closest to a dendrite.
10. The Indian python is a species of snake _____ existence is threatened.

EQUIVALENTS OF *WHOSE, WHERE,* AND *WHEN*

Some writers prefer to avoid using *whose, where,* and *when* in relative-clause sentences. Each of these words has at least one equivalent phrase. Some of these equivalents require specific changes in sentence structure.

Whose (personal) = *who has, with:*

The patient <u>whose leg</u> is broken is in room 37.
The patient <u>who has</u> a broken leg is in room 37.
The patient <u>with</u> a broken leg is in room 37.

Whose (impersonal) = *of which, that has, with:*

Pluto is a planet <u>whose surface</u> is covered with ice.
Pluto is a planet <u>the surface of which</u> is covered with ice.
 (Notice the change in subject position. This is very formal.)
Pluto is a planet <u>that has</u> an ice-covered (or icy) surface.
Pluto is a planet <u>with</u> an ice-covered (or icy) surface.

Where = *in which, at which* (or other prepositions of place + *which*):

The spleen is the organ <u>where</u> bile is produced.
The spleen is the organ <u>in which</u> bile is produced.
Zero degrees C is the point <u>where</u> water freezes.
Zero degrees C is the point <u>at which</u> water freezes.

When = *in which, during which, at which* (or other prepositions of time + *which*):

> The Industrial Revolution was a period in history <u>when</u> machines started to replace human labor.
> The Industrial Revolution was a period in history <u>in which</u> (<u>during which</u>) machines started to replace human labor.

EXERCISE 2.5 Rewrite the sentences in Exercise 2.4, using equivalents for *whose, where,* and *when.*

Example: A record player *whose* needle is worn can damage records.

> A record player the needle <u>of which</u> is worn can damage records. (very formal)
> A record player <u>that has</u> a worn needle can damage records.
> A record player <u>with</u> a worn needle can damage records.

Note that in some *whose* clauses, *whose* cannot be replaced with *that has* or *with.*

EMBEDDED QUESTIONS[3]

Embedded questions are constructed by means of sentence combining. Like relative clauses, they consist of a main clause and a subordinate clause. However, since an embedded question is a type of noun clause, the subordinate clause can be either the subject or the object of the main clause. The subordinate clause is always a transformed question that is embedded in the main clause. Consider these examples:

    SUBJECT   **VERB**   **COMPLEMENT**
How aspirin works is still not understood.   (statement)
 SUBJECT   **VERB**   **OBJECT**
Scientists do not know how aspirin works.   (statement)
   SUBJECT **VERB**   **OBJECT**
Do scientists know how aspirin works?   (question)

---

[3]The term embedded question used here actually refers to embedded WH-questions. Embedded yes/no-questions also occur. These require the addition of *if* or *whether (or not)* to the statement form of a yes/no question. Embedded yes/no-questions can function as both subjects and objects with *whether (or not)* but can only function as objects with *if.* Look at these examples:

<u>Whether or not the new comet can be seen without a telescope</u> is uncertain.
It is not known <u>whether or not the new comet can be seen without a telescope</u>.
[Embedded yes/no-question: Can the new comet be seen without a telescope?]
The patient asked <u>if salt causes high blood pressure</u>.
[Embedded yes/no question: Does salt cause high blood pressure?]

Like relative clauses, embedded questions have both an object form (O form) and a subject form (S form). The subordinate clause in an O-form embedded question can be constructed with either a clause or an infinitive phrase.

Object-Form Embedded Questions with a Clause

In an object-form embedded question, the question word is the object (or the object of a preposition) in the subordinate clause. (We know that the question word concerns the object rather than the subject because of the presence of the auxiliary verb *do* in the question. *Do* is used only with question words concerning the object.)

The doctor does not know X.	(main clause)
+	
X = Where did the patient buy the medicine?	(subordinate clause)

(1)   The doctor does not know where the patient bought the medicine.

To get sentence 1, we first change the subordinate-clause question into a statement:

X =   Where did the patient buy the medicine?
step 1:   The patient bought the medicine in ?
step 2:   The patient bought the medicine where.

Like the relative pronoun in an object-form relative clause (see Unit I, Section 3), the question word *where* must be moved to the front of the clause:

step 3:   ... where the patient bought the medicine.

Now that X is in the correct form, we substitute it in the main clause to get the finished sentence:

(1) The doctor does not know where the patient bought the medicine.

EXERCISE 2.6   Combine each of the following pairs of sentences into an O-form embedded question.

Example: (a) John does not know X.
                    (b) X = What is an isotope?
Answer: John does not know what an isotope is.

1. (a) The chemistry professor described X.
   (b) X = What is a p-orbital?
2. (a) No one can remember X.
   (b) X = When was the equipment delivered?

3. (a) The students asked X.
   (b) X = Why are radio telescopes so large?
4. (a) X is explained in the feasibility study.
   (b) X = How much does the project cost?
5. (a) Physicists and astronomers are trying to determine exactly X.
   (b) X = When did the world begin?
6. (a) A meteorite gives us a good example of X.
   (b) X = What was the earth like when it was formed?
7. (a) Low-energy accelerators are used to learn X.
   (b) X = How do particles act in the nucleus?
8. (a) Heat moves slowly, spreading out gradually by X.
   (b) X = What do we call diffusion?
9. (a) X-ray technicians located precisely X.
   (b) X = Where was the blood clot?
10. (a) Biologists had not predicted X.
    (b) X = How destructive would the Dutch elm fungus be?

Object-Form Embedded Questions with an Infinitive

Object-form embedded questions can also be constructed with an infinitive phrase. In sentence 1, the main subject, *doctor,* is different from the clause subject, *patient:*

(1) The doctor does not know where the patient bought the medicine.

If the main subject and the clause subject are the same (or if a general subject such as *people* is meant), we can use (1) a subject plus a modal verb (see sentence 2a) or (2) an infinitive verb in the embedded question (see sentence 2b). The subject plus modal verb usually indicates **one time**. An infinitive, because it has no tense, indicates **general time,** although it often implies a future event.

The nurse knows X.

\+

X = What should the nurse do?

(2a) The nurse knows what she should do.   (one time)
(2b) The nurse knows what to do.   (general time)

Sentence 2a indicates that the nurse knows what she should do in this particular case. Sentence 2b indicates that the nurse knows what to do in general (in other words, she is well trained and competent) and therefore what to do now and in the future.

EXERCISE 2.7   Combine the following into O-form embedded questions with an infinitive phrase to show general time.

Example: (a) The students do not know X.
             (b) X = What should the students read?
Answer: The students do not know what to read.

1. (a) Your lab partner will tell you X.
   (b) X = When should you stop adding water?
2. (a) Learning X takes a lot of practice.
   (b) X = How can writers write concisely?
3. (a) The film shows patients X.
   (b) X = How can patients reduce high blood pressure?
4. (a) A bibliography suggests X.
   (b) X = Where can you look for further information?
5. (a) How do hibernating animals know X?
   (b) X = When will they wake up?
6. (a) The pharmacist knows X.
   (b) X = How much acetic acid should the pharmacist use?
7. (a) Because both explanations were logical, the committee did not know X.
   (b) X = Whom should the committee believe?
8. (a) X is described on the container.
   (b) X = What should people do in an emergency?
9. (a) X is common knowledge to any Asian farmer.
   (b) X = Where should one plant rice?
10. (a) The Nuclear Regulatory Commission was not sure X.
    (b) X = Whom should it blame for the radiation leak?

Subject-Form Embedded Questions

In a subject-form embedded question, the question word is the subject of the subordinate clause. (We know that the question word concerns the subject rather than the object because there is no auxiliary form of *do* in the question.)

The police do not know X. (main clause)
+
X = Who lives here?   (subordinate clause)

(3)   The police do not know who lives here.

As in the S-form relative clause, the question word *who* in sentence 3 does not need to be moved because it is already at the beginning of the subordinate clause.

EXERCISE 2.8   Combine the following into S-form embedded questions.

Example: (a) The office knows X.
             (b) +X = Who works in the lab?
Answer: (b) The office knows who works in the lab.

1. (a) No one knows X.
   (b) X = How many people are carriers of the disease?

2. (a) The officials disagreed on X.
   (b) X = Who should build the weather satellite?
3. (a) Have cancer experts determined X?
   (b) X = Which gene alterations are responsible for leukemia?
4. (a) The civil-engineering department told us X.
   (b) X = What kind of structure is best for this site?
5. (a) X is still a mystery.
   (b) X = What makes the continental plates move?
6. (a) The hospital will not say X.
   (b) X = Whose medical records were destroyed in the fire?
7. (a) Can you guess X?
   (b) X = Which element was discovered in Berkeley, California?
8. (a) The team is still trying to work out X.
   (b) X = What type of particle passed through the chamber?
9. (a) X is a matter of conjecture.
   (b) X = What preceded the "big bang"?
10. (a) Geologists have a good idea of X.
    (b) X = What forces trigger earthquakes?

Section 3
REDUCTION AND MODIFICATION

REDUCING DEFINING RELATIVE CLAUSES

Many relative-clause sentences can be made more concise through reduction. This reduction is very common in scientific writing. However, reduction is usually a choice, not a requirement.

Reducing Object-Form Defining Relative Clauses

Before we discuss the reduction of object-form (O-form) relative clauses, try the following exercise.

EXERCISE 3.1   Can you remove any words in the following sentences without changing the meaning? If so, draw a circle around those words.

1. The planet that the astronomers observed is Saturn.
2. Invertebrates are animals that have no backbone.
3. The hospital was not satisfied with the nurse whom the doctor hired.
4. The food that you eat should be varied.
5. Iodine is an element that aids in the prevention of goiter.
6. A geologist notices the manner in which strata are deposited.

You probably noticed that you can remove the relative pronouns, *that, whom,* and *that* in sentences 1, 3, and 4. Notice that these sentences are all object-form relative-clause sentences (see Unit I, Section 3). Sentences 2, 5, and 6 cannot be changed. Sentences 2 and 5 are subject-form relative-clause sentences (see Unit I, Section 3). Sentence 6 contains an object-form relative clause, but there is a preposition before the relative pronoun *which* (see "Relative Clauses With Prepositions," Section 2 of this unit).

This leads us to our first rule for reducing defining relative clauses: Remove the relative pronoun in an object-form relative clause unless it is preceded by a preposition.

EXERCISE 3.2 Keeping the same word order, remove any unnecessary relative pronouns in the following sentences.

Example: The lab that we visited was very modern.
  Answer: The lab we visited was very modern.

1. Venus is a planet that is structurally very similar to the earth.
2. The pollution that industry produces is a serious threat to the environment.
3. The zoologist with whom they studied won the Nobel Prize.
4. The computer that the EECS department installed has been down for two weeks.
5. The woman whom I met at the conference is a well-known particle physicist.
6. A chicken is a mechanism by means of which an egg produces another egg.
7. Caffeine is a chemical whose properties can affect the human heart.
8. Chicken pox is a disease that most people contract as children.
9. Metals that offer no resistance to electricity at extremely low temperatures are called superconductors.
10. The male insects that the group studied all died within a week.

EXERCISE 3.3 Reduce the sentences in Unit II, Section 2, Exercise 2.3.

Example: Latex is the substance that rubber is made from.
 Answer: Latex is the substance rubber is made from.
Analysis: Is the resulting sentence formal or informal?

Reducing Subject-Form Defining Relative Clauses

Before we discuss the reduction of subject-form (S-form) relative clauses, try the following exercise.

EXERCISE 3.4   Can you remove any words or phrases in the following sentences without changing the meaning? If so, draw a ⟨circle⟩ around these words.

1. One solvent that is used in dry cleaning is benzine.
2. The leaf that is growing on the left does not respond to light.
3. A physicist who is at the CERN laboratory noticed the particle track.
4. The only water that is drinkable here is bottled water.
5. The pump that is the closest to the venturi valve has broken down twice.
6. Venus is a planet that is covered with clouds.

There are two ways to reduce subject-form relative clauses: (1) by removing the relative pronoun + *be*, and (2) by removing the relative pronoun and changing the verb to the V_{ing} form.

Removing the relative pronoun + *be*.   Subject-form (S-form) relative clauses can be reduced by removing the relative pronoun (*who, whom, which,* and *that;* but not *whose* or the relative adverbs *where* or *when*) **plus** *be* in all cases except where only a noun phrase or certain adjectives remain. This reduction is common but not required in English. *Who, whom, which,* and *that* + *be* can be removed:

1a. if the word following *be* is a past participle (V_{ed2}) with a postmodifying phrase, as in a passive relative clause:
The plutonium <u>which was stolen from the lab</u> was never found.
The plutonium <u>stolen from the lab</u> was never found.
1b. if the word following *be* is one of a group of past participles **without** a postmodifying phrase. These special participles (a) show a specific state, (b) show a temporary state, or (c) refer to something previously mentioned:
   a. The words <u>which are underlined</u> should be changed.
      The words <u>underlined</u> should be changed.
   b. The nature of the stress <u>which is involved</u> constantly change s.
      The nature of the stress <u>involved</u> constantly changes.
   c. The curve <u>which is shown</u> presents the final results.
      The curve <u>shown</u> presents the final results.
2. if the word following *be* is a present participle (V_{ing}) with a postmodifying phrase:
The weather system <u>which is approaching the coast</u> is a hurricane.
The weather system <u>approaching the coast</u> is a hurricane.
3. if the words following *be* constitute a prepositional phrase:
A body <u>which is at rest</u> has no motion in relation to an observer.
A body <u>at rest</u> has no motion in relation to an observer.
4. if the word following *be* belongs to a special group of adjectives ending with *-ble* (e.g., *capable, possible, responsible, visible*), especially

if the antecedent of the relative pronoun is modified by a unique adjective, such as *the only* (see "Ranking Adjectives," Section 1 of this unit):

A small scar was the only change <u>that was visible</u> three weeks after the operation.

A small scar was the only change <u>visible</u> three weeks after the operation.

Reduction by removal of the relative pronoun + *be* is **not** possible:

1. if the word after *be* is a noun phrase:
   Incorrect: *There are two elements liquids at room temperature.
   Correct: There are two elements that are <u>liquids at room temperature</u>.
2. if the word after *be* is a single adjective not ending in *-ble:*
   The plastic <u>which was lighter</u> failed to support the weight.
   Incorrect: *The plastic lighter failed to support the weight.
   Correct: The <u>lighter</u> plastic failed to support the weight. (The adjective must be moved to the front of the noun.)

EXERCISE 3.5   Reduce the following S-form relative-clause sentences if possible.

Example: The nurse who was hired by the doctor was very young.
 Answer: The nurse hired by the doctor was very young.

1. The person who is responsible is the doctor who gave the injection.
2. The most distant object that is known in the universe is a quasar.
3. Stars that are red are older than stars that are blue.
4. The sentences that are underlined should be changed in the report.
5. The computer is a system that is simple enough for a child to use.
6. The liquid that is boiling in the Dewar jar is liquid helium.
7. The only theory that was possible did not account for all the data.
8. A problem that is in many pressurized-water nuclear reactors is the deterioration of steam-generator tubes.
9. Scientists who are studying volcanoes know that small earthquakes usually precede eruptions.
10. Apnea is a sleep disorder that is characterized by frequent respiratory failure.

Removing the relative pronoun and changing the verb to V$_{ing}$.   An S-form relative clause can also be reduced by removing the relative pronoun and changing the verb to the V$_{ing}$ form. Since the present participle (V$_{ing}$) has no tense, it indicates general time and therefore a fact. For this reason, this reduction is not possible if the statement describes a single event:

Single event: The scientist <u>who discovered</u> DNA was Friedrich Meischer.
Incorrect: *The scientist discovering DNA was Friedrich Meischer.
Fact: The planets <u>that circle</u> the sun constitute the solar system.
        The planets <u>circling</u> the sun constitute the solar system.

This type of S-form relative-clause reduction is **not** appropriate:

    1. if the verb phrase in the relative clause includes a modal:

Penicillin is a drug that <u>may cause</u> a strong reaction.
Incorrect: *Penicillin is a drug may causing a strong reaction.
               *Penicillin is a drug causing a strong reaction. (This
               sentence is grammatically correct but is not true.
               Penicillin does not **always** cause a strong reaction.)

    2. if the verb in the relative clause is *be* (see also Unit V, Section 2,
       "Dangling Modifiers"):

The lecture will be given by a man who <u>is</u> an electrical engineer.
Incorrect: *The lecture will be given by a man being an electrical
           engineer.

EXERCISE 3.6   Reduce the following S-form relative-clause sentences if
possible.

Example: People who travel by plane often experience jet lag.
 Answer: People traveling by plane often experience jet lag.

    1. Nurses who live in infected areas should be inoculated.
    2. The speaker at the conference was an engineer who is also a
       zoologist.
    3. The river that supplies water to the central valley contains small
       amounts of dioxin, a known carcinogen.
    4. The person who will have the answer to this program is the pro-
       ject manager.
    5. Researchers that test new undersea cables prefer optical-fiber ca-
       bles for long distances.
    6. The dam that produced the highest amount of hydroelectric
       power in 1981 was the Itaipu on the Brazil-Paraguay border.
    7. The restrictions apply only to researchers who do recombinant-
       DNA experiments.
    8. The physicists who discovered radium and polonium in 1898 were
       Marie and Pierre Curie.

9. The animal that can run faster than any other is the cheetah.
10. Air-pollution emissions that are high in sulfur oxides come mostly from electric utilities.

EXERCISE 3.7   Reduce the relative clauses in the following passage where possible.

THE SUN[4]

Scientists have long known that the sun is a star which consists of hydrogen, helium, and seventy-two other elements. A thermonuclear furnace which is at the core converts hydrogen to helium, releasing heat and light. In recent years—largely thanks to Skylab's solar telescope—scientists have learned that the surface of the sun seethes with bubbles and eddies. The kinetic energy which is contained in this movement produces enormous noise, but because space carries no sound we cannot hear it. And there are solar "prominences" which were first spotted during eclipses. The great fingers of fire which leap thousands of miles are explained as lines of magnetic force which connect regions of opposite polarity.

EXERCISE 3.8   Write the correct active (V_{ing}) or passive (V_{ed2}) form of the verbs in parentheses.

Example: People (live) _____ near airports often suffer from heart problems (cause) _____ by noise pollution.
Answer: People living near airports often suffer from heart problems caused by noise pollution.

1. Interferon is a protein (synthesize) _____ in an animal cell as a result of infection by a virus.
2. Waste (contain) _____ radioactive materials will be stored deep in salt deposits.
3. Blood (take) _____ during a blood test is placed in a test tube (label) _____ with the patient's name.
4. Water (pass) _____ through a sewage-treatment plant is constantly monitored for microorganisms.
5. A cable (make) _____ of glass fibers transmits telephone data more efficiently than a conventional copper cable.
6. The paint (cover) _____ a metal surface protects it from corrosion.
7. The dermis of the skin is a connective-tissue layer (compose) ____ mostly of collagenous fibers.

[4]Reprinted with permission—The Toronto Star Syndicate.

8. Human infants (weigh) _____ six kilograms require an average of 690 kilocalories per day.
9. Silicate dust (surround) _____ the sun forms a ring two million kilometers wide.
10. Latex (mix) _____ with carbon and (heat) _____ with sulfur is the basis of commercial rubber.

ADJECTIVES WITH -ing OR -ed ENDINGS

Many adjectives in English are derived from verbs. Adjectives based on active verbs are formed with -ing. Adjectives based on passive verbs are formed with -ed (or with the past-participle forms of irregular verbs—see Appendix A).

Adjectives With -ing

Adjectives with -ing are commonly used with objects, materials, or systems designed to perform a certain activity:

a drilling platform   cleaning agents
a cooling tower   a cataloging system
polarizing lenses   a marketing plan

They also occur with objects or materials that act or produce an action by themselves:

a rotating star   ionizing radiation
a bleeding ulcer   superconducting metals

Adjectives With -ed

Adjectives with -ed describe an object or material to which something has been done:

a labeled test tube   an exploded view
a machined finish   distilled water

They also describe inherent attributes or characteristics that may or may not imply that a process has been undergone:

a colored liquid   an inherited characteristic
a disc-shaped cell   chlorinated hydrocarbons

Adjectives With Both -ing and -ed

Adjectives usually occur exclusively with one ending or the other unless they are derived from process verbs (to be discussed shortly). However, a few adjectives occur with both endings:

a polishing machine = a machine that polishes
a polished machine = a machine that somebody has polished.

a connecting rod = a rod that connects a piston to the crankshaft
a connected rod = a rod that somebody has connected.

EXERCISES 3.9   Write the correct -ing or -ed adjective form of the verbs in parentheses.

Example: Only (sterilize) instruments may be used.
 Answer: Only sterilized instruments may be used.

1. The (build) _____ code states that the foundation of this building must be twenty-five feet deep.
2. An aneroid barometer is based on an (evacuate) _____ container.
3. Many display devices use light-(emit) _____ diodes.
4. The (grind) _____ teeth of modern orangutans use a mortar-and-pestle (chew) _____ method.
5. A weather satellite supplies (detail) _____meteorological data.
6. The (underlie) _____ principle of fractal-modeled physical processes is still not understood.
7. A new cholesterol-(lower) _____ drug has been found to reduce heart disease.
8. Some (inject) _____ medications cause serious side effects.
9. Oil is usually found in (stratify) _____ rock formations.
10. The fluctuation was due to a (sample) _____ error rather than to a significant anomaly.

Adjectives Formed From Process Verbs

Adjectives formed from verbs indicating a process have a slightly different meaning that is related more closely to verb tense. The -ing adjective indicates an action in process whereas the -ed adjective indicates a completed action:

a developing cell = a cell in the process of developing
a developed cell = a mature cell that no longer grows
melting snow = snow in the process of melting
melted snow = snow that has melted and become a liquid

EXERCISE 3.10   Write the correct *-ing* or *-ed* adjective form of the verbs in parentheses.

Example: (Burn) _____ flesh does not heal quickly.
  Answer: Burned flesh does not heal quickly.

1. A (grow) _____ plant needs light, minerals, and moisture.
2. A (grow) _____ man weighs an average of seventy kilograms.
3. Space travel requires the application of (advance) _____ mathematics.
4. An (advance) _____ glacier grinds down the rocks in its path.
5. The (infect) _____ virus has still not been identified.
6. (Infect) _____ wounds must be cleaned with a germicidal solution.
7. Rockets use (compress) _____ fuel.
8. Diamond and oil are the products of the (compress) _____ action of millions of tons of rock over time.
9. (Oxygenate) _____ blood flows to the left auricle of the heart.
10. The (oxygenate) _____ process takes place in the lungs.

Section 4
VERBS: THE PASSIVE VOICE

About one third of all verbs in scientific writing occur in the passive form. First we will look at the structure of the passive, and then we will discuss its use.

STRUCTURE OF THE PASSIVE

The passive is a grammatical structure that allows the object of a verb to be placed in the subject position. Why do we need to put the object in the subject position? There are several reasons.

The subject position is the strongest position in an English sentence. Therefore, if a weak noun—one with little information or one that is obvious to the reader—is used as a subject, this strong position is wasted. Look at this active sentence:

SUBJECT VERB OBJECT ADVERBIAL

(1) People speak English in London.

This sentence is grammatically correct, but the subject is weak because we know that only **people** speak languages. This sentence is much better in the passive. To change an active sentence into a passive sentence, we move the object to the subject position and change the verb to a form of *be* plus the past participle of the verb:

(1) People   speak   English in London.   (active)

(2) English is spoken   in London.   (passive)

Two processes are involved in the transformation of the active verb: (1) selecting the correct form of *be* and (2) changing the active verb to the past participle (in the example above, *speak* ⟶ *spoken*).

      In sentence 1, the active verb *speak* gives us two very important pieces of information: the tense of the verb (simple present in this case) and the number (plural in this case). In the passive verb form, this information must be transferred to the verb *be* because the past participle does not show tense or number:

TENSE
Simple Present

(1) *to speak* ⟶   *speak*

(2) *to be* ⟶   *is/are*   +   *spoken* (past participle of *speak*)

      The number of the verb is always controlled by the noun that occupies the subject position (see Unit I, Section 4, "The Agreement Between Subjects and Verbs"). In sentence 1, the subject is the plural noun *people*. Therefore, the verb has the plural form *speak*. However, in sentence 2, the subject is the singular noun *English*. Therefore, the form of *be* must also be singular.

NUMBER
Singular

(1) Passive subject (active object)⟶   *English*

(2) *to be*   ⟶   *is*

Let us review the form of the passive in the five verb tenses most commonly used in scientific writing:

1. Simple Present:
   Active   People <u>speak</u> English in London.
   Passive   English <u>is spoken</u> in London.
2. Simple Past:
   Active   People <u>spoke</u> English in London in the 1400s.
   Passive   English <u>was spoken</u> in London in the 1400s.
3. Simple Future:
   Active   People <u>will speak</u> English in London forever.
   Passive   English <u>will be spoken</u> in London forever.
4. Present Perfect:
   Active   People <u>have spoken</u> English in London for years.
   Passive   English <u>has been spoken</u> in London for years.
5. Present Continuous:
   Active   People <u>are speaking</u> English in London right now.
   Passive   English <u>is being spoken</u> in London right now.

Notice in the present-continuous passive that the continuous *-ing* aspect cannot be added to *is* because we cannot say *"ising"* in English. For this reason, we add a second form of *be (being)* to carry this information.

EXERCISE 4.1   Add the correct form of the passive auxiliary *be* to the following sentences.

Example: Glass _____ made from sand.
 Answer: Glass is made from sand.

1. Aluminum _____ produced from bauxite.
2. The first locomotives _____ powered by steam.
3. Hundreds of people _____ killed because of storms this winter.
4. A lot of coffee _____ grown in South America.
5. The solar system _____ formed approximately 4.5 billion years ago.
6. The brain chemical that regulates growth _____ recently synthesized.

7. The next space module _____ sent to Venus.
8. The pores in a leaf _____ called stomata.
9. Microcomputers _____ used more and more in the future.
10. A pound of potatoes _____ needed for the experiment.

CHOOSING THE PASSIVE OR THE ACTIVE VOICE

Changing a sentence from the active to the passive voice or vice versa is rare in English discourse. Of course, we must know how to do this to correct an incorrect choice of form. It is much more important to be able to recognize **when** a subject requires the active voice and when it requires the passive.

An active verb form follows an active, dynamic subject in most cases. An active subject usually **acts:** it causes something to happen or does something to the object that receives that action:

SUBJECT⟶ ⟶OBJECT
The astronauts sent a message.

The passive verb form usually follows a passive, receptive subject: it is the effect of something that happens. Sometimes, if it is not too weak, the active subject is included in a *by*-phrase. It is then called an agent:

SUBJECT⟵ ⟵AGENT
A message was sent by the astronauts.

EXERCISE 4.2 Choose the correct active or passive form of the verb in parentheses. Be sure the verb is in the correct tense.

Example: In the Stone Age, tools (make) of stone.
Answer: In the Stone Age, tools were made of stone.

1. Friction (reduce) by the application of oil.
2. The earthquake (occur) on April 18, 1906.
3. Fluorescent lighting (link) to melanoma, a vicious form of skin cancer, in the last few years.
4. The next space mission (map) Venus by means of radar.
5. X rays (discover) by Roentgen in 1895.
6. Now, several patients (test) for immune-deficiency diseases.
7. Heating the sample (result) in the generation of oxygen gas.
8. The light that you see here shows that the laser beam (reflect) from the moon back to the earth at this moment.
9. Physicists (recently find) traces of the subatomic particle known as W.
10. The ejection of volcanic dust into the atmosphere (slightly lower) global temperatures for the next few years.

11. Evidence that the whole universe rotates (present) at the symposium last week.
12. Weather forecasters claim that the weather (improve) next year.
13. A microscope (usually compose) of an objective, a specimen stage, and a light source.
14. Electronic engineers (currently investigate) the feasibility of three-dimensional television.
15. Thanks to satellite photographs, the earth's aurora (show) to be nearly round.

PASSIVE STRUCTURES WITH *BY*-AGENTS

Approximately 80 percent of all passive sentences in English do not include the active subject. In the remaining twenty percent, many active subjects are included in a *by*-phrase. The object of the preposition *by* is called a *by*-agent.

At the beginning of this section we used the passive verb form because the subject was weak. A weak subject is never included in a *by*-phrase:

Incorrect:
*English is spoken in London by people.
*The bridge was built in 1934 by somebody to reduce traffic on other roads.

However, a passive sentence can sometimes give us useful information by answering the question *By whom?* or *By what?* For example,

This report was requested <u>by the Lightman Chemical Company</u>. (The Lightman Chemical Company requested this report.)
Decayed teeth are removed <u>by dentists</u>.
The moon is held in orbit <u>by the earth's gravity</u>.
Food is digested <u>by enzymes in the stomach</u>.

EXERCISE 4.3   Complete the following sentences with the *by*-agent in parentheses **if necessary.**

Example: Poisons are removed from the blood. (the liver)
  Answer: Poisons are removed from the blood by the liver.

1. Milk for human consumption is produced. (cows)
2. Gunpowder is used in making weapons. (the weapons industry)
3. The Golden Gate Bridge was built in 1937. (workers)
4. Some animals use trees to build dams. (beavers)
5. The space shuttle was first launched in 1981. (NASA)
6. A pearl is produced in response to an irritation. (oyster)
7. In plants, water is absorbed. (the roots)
8. The doctor requested that an X ray of the patient be taken. (nurse)

9. The structure of the DNA molecule was discovered in 1953. (scientists)
10. The structure of the DNA molecule was discovered in 1953. (Watson and Crick).

A second reason to use the passive verb form is to keep the more relevant noun phrase in the subject position. For example, in a paragraph about the different devices a chemical engineer uses, this sentence might occur:

> The chemical engineer uses heat exchangers, venturi meters, and other devices to control the temperature and pressure of fluids.

However, in a paragraph about heat exchangers, this sentence might occur:

> Heat exchangers are used (by chemical engineers) to control the temperature of a fluid.

In other words, we usually put the main topic of a paragraph in the strong subject position.

EXERCISE 4.4   Underline the complete subject of each sentence in the following paragraph. If the subject is different from the topic, analyze the object to see if it is closer to the topic in meaning; if so, move the object to the subject position. Rewrite the paragraph, using pronouns to avoid redundancy where possible, and delete any unnecessary *by*-phrases.

Some Uses of Plants[5]

A plant is a living organism. Different parts, each with particular purposes or functions, make up the plant. Some parts of the plant may be removed without harming it. Food is stored in the roots of plants such as beets, carrots, and potatoes. Other plants are able to take nitrogen from the air and add it to the soil. Scientists call these plants legumes. If they are plowed under, they make the soil more fertile.

HOW-AGENTS

Agents can also be determined by asking the question *How?* The main difference between a *by*-agent and a *how*-agent is that a *how*-agent indicates purpose, intent, or a desired goal, whereas a *by*-agent simply indicates that something happened. Look at these sentences:

(1) The man was killed <u>by</u> a stone.   (*by*-agent)
(2) The man was killed <u>with</u> a stone.   (*how*-agent)

---

[5]Adapted from Alan Mountford, *English in Agriculture* (Oxford: Oxford University Press, 1977), p. 1.

In sentence 1, we understand that the man was killed because a stone fell, or he fell on a stone and hit his head. The emphasis is on the stone as the cause of death. In sentence 2, we understand that the man was killed by another person using a stone as a weapon.

How-agents are often attached to passive sentences, but unlike *by*-agents, they also occur in active sentences. *How*-agents are indicated by the prepositions *by, with,* and *by means of.* Notice that *by* occurs in both *by*-agents and *how*-agents.

by

When used with a *how*-agent, *by* must be accompanied by (1) the zero article ($\emptyset$) plus a singular countable or uncountable noun (see Unit VI, Section 1, "Idiomatic Structures With $\emptyset$") or (2) a V_{ing} phrase.

PASSIVE	ACTIVE
The equipment was delivered <u>by truck</u>.	The equipment came <u>by truck</u>.
A wind generator is powered <u>by wind</u>.	A wind generator operates <u>by wind</u>.
The patient was revived <u>by injecting him with insulin</u>.	Doctors sometimes revive patients <u>by injecting them with insulin</u>.
Water is commonly purified <u>by distilling it</u>.	Desalinization plants purify water <u>by distilling it</u>.

with

With is commonly used with tools, devices, and materials. Unlike *by*, it can never be used with $\emptyset$ plus a singular countable noun.

PASSIVE	ACTIVE
Teeth are removed <u>with forceps</u>.	Dentists remove teeth <u>with forceps</u>.
The screw was removed <u>with a screwdriver</u>.	Remove the screw <u>with a screwdriver</u>.
Pressure can be measured <u>with a venturi meter</u>.	We measured the pressure <u>with a venturi meter</u>.
Houses are often insulated <u>with fiberglass</u>.	Carpenters commonly insulate houses <u>with fiberglass</u>.
Food is digested <u>with enzymes</u>.	The stomach digests food <u>with enzymes</u>.

by means of

By means of is similar to *with* except that it emphasizes the **process** that a tool, device, or material performs.

PASSIVE	ACTIVE
Teeth are removed <u>by means of forceps</u>.	Dentists remove teeth <u>by means of forceps</u>.
Food is digested <u>by means of enzymes</u>.	The stomach digests food <u>by means of enzymes</u>.

If the *how*-agent already represents a process, *by means of* may be shortened to *by*.

PASSIVE	ACTIVE
The patient was revived <u>by (means of) an injection of insulin</u>.	Doctors sometimes revive patients <u>by (means of) an injection of insulin</u>.
Water is commonly purified <u>by (means of) distillation</u>.	Desalinization plants purify water <u>by (means of) distillation</u>.

The uses of *by*-agents and *how*-agents are summarized in the following chart:

by-AGENT (PASSIVE)	how-AGENT (ACTIVE OR PASSIVE)
by + noun phrase	*by* $\begin{cases} + \emptyset + \text{singular noun} \\ + V_{ing} \text{ phrase} \end{cases}$ *with* + noun phrase *by means of* + noun phrase *by (means of)* + process noun

EXERCISE 4.5   Fill the blanks with *by*, *with*, or *by means of*. Indicate whether the final noun phrase is a *by*-agent (by) or a *how*-agent (how). Sometimes more than one answer is possible.

Example: _____ She opened the box _____ a knife.
  Answer: <u>how</u> She opened the box with a knife.

1. _____ The lumber was transported _____ train.
2. _____ Current can be varied in a circuit _____ adjusting the rheostats.
3. _____ Earthquakes are usually caused _____ the movement of the continental plates.
4. _____ The age of the sample was determined _____ the carbon 14 dating method.
5. _____ The carburetor can be adjusted _____ a small wrench.
6. _____ An electrical connection is greatly improved _____ the application of solder.
7. _____ The boiling point of a liquid is affected _____ altitude.

8. _____ Steel surfaces can be protected _____ applying a thin coating of machine oil.

9. _____ Many electronic devices are assembled _____ hand.

10. _____ The colors of the aurora borealis are generated _____ the effect of solar particles on the various atmospheric gases.

11. _____ Many cancer patients hope to be cured _____ interferon, a protein produced by an animal in response to a virus.

12. _____ The primitive cooling earth was darkened for centuries _____ hot, dense clouds.

13. _____ The response was measured _____ receptors placed beneath the skin.

14. _____ The direction of growth of a plant is controlled _____ light (phototropism).

15. _____ The orbit was calculated _____ the formula for centripetal force, $mr\omega^2$, where $m = mass$, $r = radius$, and $\omega = angular$ velocity.

16. _____ The body temperature of a higher mammal is regulated _____ the hypothalamus.

17. _____ The television repairman checked the faulty transistor _____ an oscilloscope.

18. _____ The plants in forests and jungles fertilize each other _____ fruit and leaves that fall to the ground and enrich the soil.

Sometimes a *by*-phrase is used to express a *how*-agent that is too long and detailed to be the subject of an active sentence. This is usually the case if the agent is a principle or a method. The following active sentence with a condensed method as a subject is correct:

Active: The process of distillation purifies water.
Passive: Water is purified by the process of distillation.

The choice of the active or passive form would be determined by the topic of the paragraph.

The process of distillation consists of a series of steps: heating water to the boiling point, condensing it, and collecting the product in a container. It is common to use the V_{ing} form to describe methods, principles, or devices in detail. If we want to use this description instead of the word *distillation,* we usually use a passive sentence:

Water is purified by heating it to the boiling point, condensing it, and collecting the product in a container.

Incorrect:
*Heating water to the boiling point, condensing it, and collecting the product in a container purifies water.

EXERCISE 4.6   The following sentences have methods, mechanisms, or principles as subjects. Replace the subject with a more detailed description using the passive and a *by*-agent with V_{ing} forms.

Example: Distillation purifies water.
 Answer: Water is purified by heating it to the boiling point, condensing it, and collecting the product in a container.

 1. Nuclear power generates electrical energy.
 2. A heat exchanger controls the temperature of a fluid.
 3. Hemodialysis cleans the blood.
 4. Smelting produces steel.
 5. Flood-control projects protect valley residents.
 6. Photographic methods reproduce photographs.
 7. Spectroscopy indicates the elements in a star.
 8. X ray crystallography determines the structure of a crystal.
 9. Statistics establish a general truth.
 10. Intravenous devices feed unconscious patients.

Section 5
WRITING AIDS

NOUN-PHRASE PARALLELISM

Parallelism in English means that two or more noun phrases, verb phrases, or modifying phrases in the same sentence must have the same grammatical form, especially if they occur in a list (a series of phrases). We will concentrate on noun-phrase parallelism in this unit because you will probably need to use parallel noun phrases in constructing a plan-of-development sentence for a description of a mechanism (see Section 6 of this unit).

Types of Noun Structures

A noun structure consists of a noun and its modification. There are four types of noun structures in English: (1) a noun phrase, (2) a gerund, (3) an embedded question, and (4) a noun clause.

Type	Examples
1. noun phrase	
a. nonmodified	teeth
b. premodified	canine teeth
c. postmodified	teeth in mammals
d. pre- and postmodified	canine teeth in mammals

2. gerund (V_{ing})
   a. nonmodified   oil refining
   b. premodified   automated oil refining
   c. postmodified   oil refining in the Middle East
   d. pre- and postmodified   automated oil refining in the Middle East

3. embedded question (question word + clause or infinitive [V_{inf}]
   a. nonmodified   what
   b. premodified   (not possible)
   c1. postmodified with clause   what the geologist measured
   c2. postmodified with V_{inf} what to measure
4a. noun clause (*that* or *if* + phrase)
   a. nonmodified   that
   b. premodified   (not possible)
   c1. postmodified with clause   that the geologist measured it
   c2. postmodified with V_{inf} (not possible)
4b. noun clause (*whether* [*or not*] + clause or infinitive)
   a. nonmodified   (not possible)
   b. premodified   (not possible)
   c1. postmodified with clause   whether [or not] he measured it
   c2. postmodified with V_{inf} whether [or not] to measure it

EXERCISE 5.1   Underline the noun phrases in the following passage. Notice that some noun phrases consist of one or more smaller noun phrases.

Science and Mathematics[6]

Not only physics but every science depends on the language of mathematics. For every science requires exact measurements and the ability to use those measurements in working out problems and finding rules. Whether you are interested in splitting atoms, discovering the structure of molecules, learning what causes depressions and prosperity in business, finding new cures for disease, designing swifter airplanes, or learning the life story of the stars, you will need to know the language of mathematics in order to further your studies.

Noun-phrase parallelism is required if two or more noun phrases occur as subjects, as objects or complements, or as objects of prepositions. These noun phrases are commonly connected by (1) *and*, (2) *both . . . and*, (3) *or*, (4) *either/or*, (5) *neither/nor*, (6) *not*, or (7) *not only . . . but (also)*:

   1. He is an engineer during the day *and* a teacher at night.
   2. He enjoys *both* engineering *and* teaching.

[6]From *World of Science* by Jane Werner Watson. © 1958 by Western Publishing Company, Inc. Reprinted by permission.

3. At first, he could not decide whether engineering *or* teaching was the best profession.
4. His parents thought that he should be *either* an engineer *or* a teacher, not both.
5. When he was at the university, he thought that he would be *neither* an engineer *nor* a teacher.
6. Later, he thought that he should be an engineer, *not* a teacher.
7. Now, he is *not only* a fine engineer *but also* an excellent teacher.

EXERCISE 5.2   Fill the blank with the correct parallel form of the phrase in parentheses.

Example: The students practiced both scaling the teeth and _____
(the students polished the teeth).
Answer: The students practiced both scaling the teeth and polishing them.

1. The artificial heart was switched off after the brain and _____
_____ had failed.
(the lungs did not function)
2. Weather forecasts try to predict not only when a storm will arrive but also _____ .
(how intense is the storm?)
3. Halogenation includes chlorination and _____
_____ . (they substitute bromine for a hydrogen atom)
4. Geologists devote their time to _____
_____ and measuring the forces that created that structure. (a geologist studies the structure of the earth)
5. There is a great difference between controlling the spread of a disease _____ .
(you must eliminate the disease)
6. One theory suggests that a meteorite caused the disruption of the Cretaceous food cycle _____
_____ .
(the dinosaurs subsequently died)
7. The report was rejected because it explained _____
_____
nor what results they found. (why did the researchers do the experiment?)
8. Making theories about certain subatomic particles is easier than _____ . (can their existence be proved?)
9. The object of the mission was not just getting the probe into orbit around the planet _____
_____ . (they will land the probe on the planet)
10. The value of field research depends on the team's accuracy _____
_____ .
(they succeed in bringing back adequate samples)

DIMENSION STATEMENTS AND MEASUREMENT WORDS

Dimension statements often occur in describing a mechanism. The four most common types of dimension statements are shown in the following examples:

Type 1. The box is two feet long.
Type 2. The box is two feet in length.
Type 3. The length of the box is two feet.
Type 4. The box has a length of two feet.

Type 1

Type 1 dimension statements may be described by the pattern

SUBJECT	be	MEASURE	Measurement ADJECTIVE
The box	is	two feet	long.

It is the only type of dimension statement to use an adjective. The measurement adjectives used in type 1 are:

long	thick	high (mountains, clouds, pyramids)
wide	deep	tall (people, buildings, trees)

Type 2

Type 2 dimension statements may be described by the pattern

SUBJECT	be	MEASURE	in + $\emptyset$ + Measurement NOUN
The box	is	⸱ two feet	in length.

The measurement nouns used in type 2 consist of the noun forms of the adjectives in type 1 (except *tall*) and other nouns:

length	breadth	diameter
width	arc	intensity (of sound)
thickness	altitude	mass
depth	area	weight
height	circumference	volume

Type 3

Type 3 dimension statements may be described by the pattern

The + Measurement NOUN + *of*	SUBJECT	be	MEASURE
The length of	the box	is	two feet.

The nouns used in this structure are all those used in type 2. Additional examples are listed below. Many of these measurement nouns can be modified to produce new concepts—for example, gear ratio, moisture content, initial cost, escape velocity, atomic number.

accuracy	gravity	range
amount	growth	rate
boiling point	heat of vaporization	ratio
capacity	index of refraction	resistance
coefficient of expansion	magnitude	span
content	melting point	specific gravity
cost	number	speed
density	population	temperature
elevation	pressure	thermal conductivity
force	proportion	velocity
freezing point	quantity	viscosity
		voltage

A few measurement nouns require the preposition *between* instead of *of*:

angle between
difference between
distance between

Type 4

Type 4 dimension statements may be described by the pattern

SUBECT	*have*	*a* + measurement NOUN + *of*	MEASURE
The box	has	a length of	two feet.

The nouns used in this structure are the same as those used in type 2 and type 3 (except for those with *between*, which can only be type 3).

EXERCISE 5.3 Indicate which of the four types of dimension statements can be used to express the information below. Then make one complete dimension statement. Use a different type for each subsequent example so that you practice using all four types.

Example: the box; volume = 20 cubic centimeters
Answer: (a) types 2–4
(b) The box is 20 cubic centimeters in volume.

1. the highway; length = 1673 miles (2677 kilometers)
2. ice; melting point = 0° C (32° F)
3. Earth; escape velocity = 11.2 kilometers/second
4. Mt. Everest; height = 8848 meters (29,028 feet)

5. standard U.S. barrel; volume = 4.21 cubic feet (0.12 cubic meter)
6. Caracas; population in 1979 = 2.85 million
7. proton; mass = 1.672×10^{-27} grams
8. World Trade Center in New York; height = 110 stories (419 meters)
9. Tokyo–Honolulu; distance = 6165 kilometers (3853 miles)
10. Mindanao Deep (Pacific Ocean); depth = 11,516 meters (37,782 feet)
11. standard typing paper; thickness = 0.01 millimeter (0.000394 inch)
12. iodine (I); atomic number = 53
13. missile-guidance system; accuracy = ± 1 mile (1.6 kilometers)
14. the two hydrogen atoms in H_2O; angle = 105°
15. delta of the Indus River; width = 208 kilometers (130 miles)
16. standard sewer pipe; inside diameter = 4 inches (10.2 centimeters)
17. Golden Gate Bridge; span = 1280 meters (4200 feet)
18. an average standing grizzly bear; height = 9 feet (2.74 meters)
19. initial pressure – final pressure; difference = 33 pounds/square inch
20. thunder; intensity = 120 decibels

MODIFYING THE BASIC DIMENSION STATEMENT

All parts of the basic dimension statement—the subject, the verb, the measure, and the measurement noun—may be modified. The only exception is the measurement adjective in type 1, which cannot be modified in any way.

Modifying the Subject

The subject in all four types of dimension statements may be modified by either premodification or postmodification:

Premodification: The <u>metal</u> box is two feet long.
Postmodification: The box <u>on the truck</u> is two feet long.
Pre- and Post-
  modification: The <u>metal</u> box <u>on the truck</u> is two feet long.

Modifying the Verb

The verb in all four types of dimension statements may be replaced with a modal auxiliary verb, particularly in specifications (see Unit VI, Section 4, "Obligation"):

Type 1:   The box ⎫ ⎧ shall be   ⎫   ⎧ two feet long.
Type 2:   The box ⎬ ⎨ should be ⎬   ⎨ two feet in length.
Type 3: The length of the box ⎭ ⎪ must be   ⎪   ⎩ two feet.
                                ⎪ will be   ⎪
                                ⎩ is to be   ⎭

Type 4:   The box $\left\{\begin{array}{l}\text{shall have}\\\text{should have}\\\text{must have}\\\text{will have}\\\text{is to have}\end{array}\right\}$ a length of two feet.

Although *be* is the most common verb in type 3 statements, it can be replaced with other verbs, such as *to equal, to be equal to,* or *to average* (if more than one subject is implied):

>   The length of the box *equals* two feet.
>   The length of the box *is equal to* two feet.
>   The length of the boxes *averages* two feet.

Modifying the Measure

The actual measure given in type 1, type 2, and type 3 dimension statements may be qualified as to its exactness. Some common qualifying words and phrases are:

precisely, exactly	The box is <u>exactly</u> 2.131 feet long.
just under, just over	The box is <u>just over</u> 2.1 feet long.
about, approximately, around	The box is <u>about</u> two feet in length.
under, over	The box is <u>over</u> two feet in length.
(somewhere) in the neighborhood of	The length of the box is <u>(somewhere) in the neighborhood of two feet</u>.
roughly	The length of the box is <u>roughly</u> two feet.

Modifying the Measurement Noun

The measurement noun is usually modified only in type 3 and type 4 statements. (Of course, if you qualify the measure in type 3, it would be redundant to qualify the measurement noun too.)

precise, exact	The box has a <u>precise</u> length of 2.131 feet.
approximate	The box has an <u>approximate</u> length of 2.1 feet.
rough	The box has a <u>rough</u> length of two feet.

EXERCISE 5.4   Choose ten of the dimension statements from Exercise 5.3. Round off the number where possible and rewrite the statement, using one of the qualifying phrases above. Be sure to use all four types of dimension statement.

Example: the highway; length = 1673 miles
 Answer: (a) Round off 1673 to 1670 or 1700.
           (b) The highway is just over 1670 miles in length.
           or The highway has an approximate length of 1700 miles.

PART II: Writing a Description of a Mechanism

Section 6
PREWRITING ACTIVITY: DETERMINING PARTS

A description of a mechanism is used in scientific writing to describe a machine, a living organism, or a system through analysis of its parts. The following exercise will help you to analyze the parts of different mechanisms.

EXERCISE 6.1 With another student, choose two of the mechanisms from the list on page 76. (1) On a piece of paper, list all the parts that you can think of for each mechanism. Compare your answers with those of your partner. (2) With your partner, classify all the parts of each mechanism into three or four major groups. Give each group a name. (3) Decide which parts are not really necessary for the operation of the mechanism and delete them.

Example: a pencil sharpener

Parts		
handle	mounting bracket	plastic-knob rivet
grinders	hole-size adjuster	manufacturer's label
shaft	gear ring	gear frame
waste container	screws	

Groups ([ ] = delete)		
(1) Frame	(2) Sharpener	(3) Driving Mechanism
mounting bracket	hole-size adjuster	handle
[screws]	grinders	shaft
gear frame		gear ring
waste container		[plastic knob]
[manufacturer's label]		[rivet]

a.	a camera	g.	a voltmeter	m.	human skin
b.	a microscope	h.	a relay	n.	a roller bearing
c.	an electric	i.	a vacuum tube	o.	a pump
	motor	j.	the heart	p.	an air compressor
d.	a tree	k.	an eye	q.	a paramecium
e.	a loudspeaker	l.	an oil-	r.	the feeding mechanism of
f.	a volcano		distillation		a flea
			column		

Section 7
STRUCTURE

A typical description of a mechanism introduces the subject, describes each part in detail, and makes a conclusion. If a mechanism has three parts, the description usually consists of five paragraphs: one for the introduction, one for each part, and one for the conclusion. Of course, more paragraphs would be required if there were a high degree of detail.

PARAGRAPH 1: INTRODUCTION

The introduction to the description of a mechanism consists of three to four sentences. They should answer the following questions, in this order:

1. Formal definition: What is it?
2. Purpose: What is it for?
3. External description: What does it look like?
4. Plan-of-development sentence: What are its parts?

Formal Definition

A formal definition (see Unit I, Section 2, "The Formal Definition") is the best way to begin a description of a mechanism because it brings the reader directly to the point. In scientific English, we do not use opening phrases such as "I am going to tell you about X" or "X is very important."

Purpose

If the purpose of the mechanism is not stated in the formal definition, it can be expressed with sentences such as these:

Example: a soldering iron
The (main) purpose of a soldering iron is to melt solder.
Soldering irons are used (mainly) to melt solder.
Soldering irons provide the heat necessary to melt solder.

Example: the adrenal gland
The function of the adrenal gland is to secrete adrenalin.
The adrenal gland secretes adrenalin.

External Description (Optional)

Many descriptions of mechanisms tell the reader what the mechanism looks like on the outside, often for identification purposes. An analogy is often used to do this.

Examples:
A cyclotron has the shape of a doughnut (or torus).
A lymph node is oval or round.
The vacuum-advance device looks like a tiny flying saucer.

Plan-of-Development Sentence

The plan-of-development sentence is the most important sentence in the entire description. It tells the reader the order of presentation of the different parts. There are several ways to write this sentence:

X consists of A, B, and C.
X is comprised of A, B, and C.
X is made up of A, B, and C.
X has the following parts: A, B, and C.

Example:
A pencil sharpener consists of
A pencil sharpener comprises
A pencil sharpener is made up of
A pencil sharpener has the following parts:
} { a frame, a sharpener, and a driving mechanism.

The plan-of-development sentence should not contain more than four parts. We do not want sentences that look like this:

Incorrect:
*A microscope consists of the eyepiece, the ocular tube, the coarse-adjustment knob, the fine-adjustment knob, the objective lenses, the stand, the stage, the condenser, the condenser adjustment, the mirror, and the base.

If the mechanism contains more than four parts, the parts should be grouped into sections:

A microscope consists of an ocular section, a specimen stage, and a light source.

EXERCISE 7.1   Look at the parts (or groups of parts) you determined in Exercise 6.1. Form them into a plan-of-development sentence using one of the model sentences above.

Example: A pencil sharpener consists of a frame, a sharpener, and a driving mechanism.

PARAGRAPH 2: DESCRIPTION OF PART A

The second paragraph in a description of a mechanism describes the first part presented in the plan-of-development sentence. The opening sentence identifies that part (What is it?). The second sentence describes its purpose (What is it for?) The third and subsequent sentences describe the details of the part, including:

1. its shape
2. its size
3. its location in relation to other parts
4. how it is connected to other parts
5. what it is made of
6. what kind of finish or surface appearance it has (e.g., polished, painted, oiled).

If the plan-of-development sentence concerns groups of parts, the third sentence of paragraph two is often a plan-of-development sentence listing the parts that belong to the first group. Then, each part must be identified and described in detail (shape, size, etc.) within the paragraph.

PARAGRAPH 3: DESCRIPTION OF PART B

Paragraph 3 is constructed in the same manner as paragraph 2.

PARAGRAPH 4: DESCRIPTION OF PART C

Paragraph 4 is constructed in the same manner as paragraph 2.

PARAGRAPH 5: CONCLUSION

Most formal compositions have a concluding paragraph. In a description of a mechanism, there are several ways to write a conclusion:

1. Describe the mechanism in action.
2. Describe the advantages of the mechanism.
3. Describe the disadvantages of the mechanism.
4. Describe the special uses and/or applications of the mechanism.
5. Describe the most recent developments in the mechanism, the latest model of the mechanism, or the most recent discovery about the mechanism.

Method 5 gives the reader a "push to the future," which is a good way to end a paper. A bad way to end a paper is to use a statement such as "Now you know how a pencil sharpener works" or "The pencil sharpener is a very important device."

Section 8
MODELS

MODEL 1: A ONE-INCH MICROMETER[7]

A one-inch micrometer (Figure 2.3) is a hand device for making outside-diameter measurements from 0.000 inch to 1.000 inch. Its basic purpose is to measure small objects which require accuracy up to 0.001 inch. The main parts are the frame, a 3/8-inch tube fastened to the frame, a 1/4-inch finely machined rod, and what is called a thimble attached to the rod's threaded portion by a screw.

Figure 2.3   One-Inch Micrometer

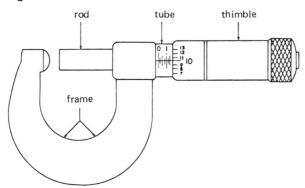

The frame is made of stainless steel and machined to a fine finish. The head, which also has a stainless, finely finished surface, mounts on the frame (Figure 2.4).

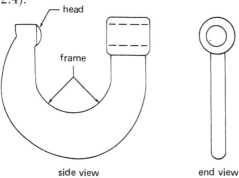

Figure 2.4   Frame

[7]Steven E. Pauley: *Technical Report Writing Today.* Copyright © 1973 by Houghton Mifflin Company. Used with permission.

The 0.375-inch-diameter stainless steel tube (Figure 2.5) attaches to the frame. A 0.250-inch-diameter hole runs through the tube and the frame, allowing insertion of a rod through the tube. Internal threads (46 per inch) on the inside of the tube are very important to the accuracy of the micrometer and will be discussed with the threaded portion of the rod. One inch of the tube is marked to indicate divisions of 0.025 of an inch. the 0.100, 0.200 and 0.300, etc., are marked by 1, 2, 3, etc., to zero, which indicates 1.000 inch.

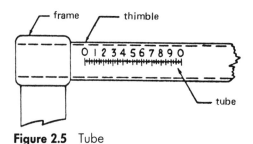

Figure 2.5 Tube

The rod, made of finely machined stainless steel, has threads on the back half to match the internal threads of the tube. The threads provide the micrometer's greatest accuracy. A complete turn of the rod equals 0.025 of an inch toward or away from the head on the frame, depending on the direction it is turned (Figure 2.6).

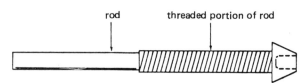

Figure 2.6 Rod

A stainless steel thimble (Figure 2.7) is attached over the threaded portion of the rod by a small screw. Markings divide the thimble's circumference into 25 increments, each of them representing 0.001 of an inch. The thimble has a knurled finish on the end for easy handling with the thumb and index finger.

Figure 2.7 Thimble

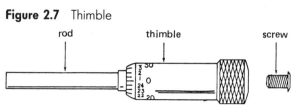

MODEL 2:THE SKIN[8]

The skin, or integument, forms a protective waterproof covering for the entire body. It is composed of two layers of epithelial and connective tissues. The outer, or surface, layer is the epidermis, which contains several layers of stratified epithelial cells. The thicker, deeper connective tissue layer is the dermis.

The epidermis is the outer, thinner layer of the integument. It consists of two or four zones of stratified squamous epithelium with increasing amounts of the protein keratin in the outermost layers. This arrangement decreases excessive water loss from the skin surface and makes the body insensitive to minor abrasions or injuries. Furthermore, the epidermis lacks blood vessels and has a limited distribution of nerve endings so that one can shave off several layers of cells without blood loss or pain.

The dermis, or corium, is a broad, dense connective tissue layer composed mostly of collagenous fibers with some elastic and reticular fibers. It contains the blood and lymph vessels, nerves, parts of the sweat and sebaceous glands, and hair roots. The dermis consists of two zones: (1) an outer papillary layer which fits against the epidermis and (2) an inner reticular layer that blends with the underlying subcutaneous tissue. The boundary between the zones is indistinct. The papillary zone is less compact, has more elastic and reticular fibers, and occurs in folds or ridges (dermal papillae) that interdigitate with the epidermis. Dermal ridges are responsible for fingerprint patterns. The compact reticular zone is composed largely of collagenous fibers running in rows parallel to the skin surface.

The skin performs many functions in the body. As a covering of the entire body, it offers protection. Its numerous sensory receptors keep us aware of the external environment and its blood supply aids in temperature regulation.

Section 9
ANALYSIS

The basic structure of a description of a mechanism, discussed in detail in Section 7, is as follows:

<u>   Title   </u>

I. Introduction
   A. Formal definition
   B. Purpose
   C. External description
   D. Plan-of-development sentence
II. Description of part A
   A. Definition
   B. Purpose

[8]L. L. Langley, Ira Telford, and John B. Christensen, *Dynamic Anatomy and Physiology,* 5th Ed. (New York: McGraw-Hill, 1980), pp. 66–68.

52
53

C. Possible details:
   1. shape
   2. size
   3. location
   4. method of attachment
   5. material
   6. finish
III. Description of part B
IV. Description of part C
V. Conclusion
  Possible concluding ideas:
  A. Mechanism in action
  B. Advantages
  C. Disadvantages or limitations
  D. Special uses or applications
  E. Latest developments or models

EXERCISE 9.1   With another student, preferably one who is in your field, analyze Model 1 in Section 8. Determine the function of each sentence in the model (formal definition, plan-of-development sentence, type of detail, type of conclusion, etc.), and label each sentence in the margin. Discuss your results with your partner. Do you agree?

EXERCISE 9.2   Analyze Model 2 in the same way as in Exercise 9.1. Compare your analyses with that of Model 1. Are they different? If so, how are they different?

Section 10
CHOOSING A TOPIC

Your assignment is to write a description of a mechanism in your own field. First you must choose a topic.[9] Many examples have been given in this unit that are good topics for a description of a mechanism (e.g., a camera, a volcano, an electric motor). Here are some additional topics related to certain majors:

Biology/Plant Pathology	Civil/Mechanical Engineering
An Angiosperm	A Cable Bridge
The Cryoscope	A Centrifugal Pump
The Feeding Mechanism of a Flea	A Hydraulic Jack
An Increment Borer	A Petrographic Microscope
	A Portable Air Compressor

[9]*Note to teacher:* Give students as much help in deciding on a topic as you find necessary. Use resources such as magazines, texts, the library, and radio and television programs. If you can arrange a visit to a place where various mechanisms are used (for example, a laboratory, an industrial plant, or a school science department), so much the better. You might want to encourage students in similar fields to make rough outlines, exchange them with each other, and discuss their strengths and weaknesses (using you as a resource person) before they begin their assignment.

Electric Engineering/Computer Science	Chemistry/Chemical Engineering
An Alternator	A Buret
A Disk Drive	An Evaporator
A Dry Cell	A Magnetic Stirrer
The Personal Computer	A Melting-Point Apparatus
	A Rotary-Drum Filter

Medicine/Physiology	Physics/Astronomy
A Blood Cell	An Electromagnet
The Cornea	The Michelson Interferometer
The Heart	A Particle Accelerator
The Thyroid Gland	The Structure of a Star

EXERCISE 10.1   Write a five-to-six-paragraph description of a mechanism, using the structure shown in Section 9. (Some of you will be able to write a composition from your own knowledge. Others will have to use the library and books or journals in their fields. Ask your teacher if you need help with reference materials.)

UNIT III
The Description
of a Process

PART I
Grammar

Section 1
ARTICLES: THE EFFECT OF MODIFICATION
ON ARTICLE CHOICE

ARTICLES WITH PREMODIFIED NOUN PHRASES

A noun phrase includes a noun plus its modification. Premodification consists of the words that come **before** the noun. These words occur in the following order:

1. Predeterminers: quantifiers such as *all, both (of), half (of), twice, a quarter of, 10 percent of*
   The predeterminers are sometimes combined with *all*:

   <u>97.2 percent of all</u> our water is in the oceans.
   <u>Half of all</u> cancers are caused by pollutants.

2. Determiners: the articles *(a, an, the, ∅)*, the words *no, this/that, every/ each, either/neither, some/any,* and possessive nouns and pronouns *(his, Mary's)*

3. Postdeterminers: cardinal numbers *(1, 2, 1000)*, ordinal numbers *(first, second, thirty-first)*, and *other* (see Unit IV, Section 3, "*Other, another*, the *others*, etc.")

4. adjectives: *black, large, complex*, etc.

1   2   3   4

Examples: Both of these two white powders contain $CaCO_3$.

1   2   3   4

Thirty percent of the lab's first published results were discarded.

We are most interested in the determiners because the articles belong to this group. A noun phrase can contain only one determiner. Look at these noun phrases:

an element	neither element
the element	this element
some elements	every element
no elements	Curie's element

When only one word can be used in this manner, we say that those words are mutually exclusive. Determiners in English are mutually exclusive. The correct determiner is always controlled by the noun except when a ranking adjective (see Unit II, Section 1) occurs in the noun phrase.

EXERCISE 1.1   Correct the following sentences by crossing out any unnecessary words.

1. Neither no this element has a lower atomic weight than hydrogen.
2. Pigeons use some the their magnetite in their skulls for navigation.
3. A lizard can see with its the each eye independently.
4. The that an Alfred Nobel's "dynamite" was invented in Sweden in 1867.
5. A glacier acts like both a this some plow and a this some file at the same time.
6. All the our any dinosaurs disappeared before the Paleocene period.
7. This each a bone is considerably larger than all that the one.
8. Their some the petroleum-producing countries will decrease their some the output next year.
9. The total volume of all the every these rivers in the world is about 300 cubic miles.
10. In a few years, only about the 25 percent of the Europe's cars will be able to carry six passengers.

The adjectives (group 4 above) usually have no effect on the choice of article. For example, in the sentence

This battery requires *(0)* water.

the zero article *(0)* is used because *water* is first-mention and general (any *water* can be used). If we add an adjective to this sentence,

This battery requires <u>distilled</u> water.

the article is still zero *(0)* because *distilled water* is still first-mention and general (any *distilled water* can be used). In Unit II, Section 1, we compared first and second mention to a distance versus a closeup photograph. The addition of an adjective to a noun phrase does not change this relationship. In other words, the first-mention (distant) view of *water* means *any or all possible water*. In the same way, the first-mention (distant) view of *distilled water* means *any or all possible distilled water*.

EXERCISE 1.2   Correct the following sentences for first-mention article use. Every sentence contains at least one error.

1. Rain forests studied were all fairly close to the equator.
2. The ocean water contains about 3.5 percent salt.
3. Dry-cleaning establishment found it difficult to remove engine-oil stains from Dr. Smith's lab coat.
4. The anodized aluminum is black in color.
5. It is common for the moisture-laden winds to lose their moisture as they move over the high mountains.
6. In general, the warm air rises and the cool air sinks.
7. African lion has an average life-span of ten years.
8. The superheated steam is widely used to drive turbines.
9. The engineers could not remove contaminated water from the Three Mile Island nuclear reactor for several months.
10. The deciduous trees are those that lose their leaves every year.

EXERCISE 1.3   Add the adjective in parentheses to the sentence.

Example: (heated) Water is often cooled by a cooling tower.
 Answer: Heated water is often cooled by a cooling tower.

1. (magnetized) A compass makes use of a needle.
2. (high-altitude) Winds produce low-pressure cells near the North Pole.
3. (clinical) The thermometer is an Italian invention.
4. (natural-habitat) Many zoos have created areas for their collections.
5. (individual coral) A polyp is basically a tube with tentacles.

6. (giant) The tortoise has an average speed of 0.17 mile per hour.
7. (infectious) Patients are usually placed in a separate ward.
8. (evergreen) Trees do not lose their foliage in winter.
9. ("permanent") Antifreeze for automobiles is composed mostly of ethylene glycol.
10. (continental) The movement of the plates has been well charted.

ARTICLES WITH POSTMODIFIED NOUN PHRASES

Postmodification consists of the words or phrases that come after the noun in a noun phrase. These are most commonly relative clauses or prepositional phrases, especially *of*-phrases. Some examples are:

Relative clause: the man (whom) we met
Prepositional phrase: the medicine on the tray
Of-phrase: the melting point of iron

EXERCISE 1.4   Underline the postmodifying phrases in the following paragraph. Draw an arrow to the word this phrase modifies.

Example: The sample in the beaker contains dioxin.
 Answer: The sample in the beaker contains dioxin.

MEASURING WITH THE STARS[1]

Scientists are trying to measure the relative motions between all the continents. Their observations involve the measurement of the distances to approximately twenty quasars. The quasars, which are located at the edge of the observable universe, are powerful sources of radio and light emission. The quasar observations depend on a technique known as Very Long Baseline Interferometry. Scientists working at widely separated land stations measure very slight differences in the arrival times of individual radio waves from the quasars to determine the relative motions of the continents.

Unlike premodification, postmodification does affect the choice of article. Postmodification usually makes the noun definite (second-mention) without first mention, just like ranking adjectives and shared knowledge, which we discussed in Unit II, Section 1.

Articles With Relative Clauses and Prepositional Phrases

Before we discuss the effect of relative clauses and prepositional phrases on the article, try the following exercise.

---

[1]Adapted from Walter Sullivan, "The Number of Inches From Here to Eternity," *The New York Times*, as reprinted in *The San Francisco Chronicle*, July 17, 1983.

EXERCISE 1.5   Add first-mention *a(n)*, *the*, or *∅* to the following blanks. Underline the relative clause or prepositional phrase in each case.

1. _____ force that the earthquake generated was measured with a seismograph.
2. In this slide, _____ nucleus of the cell is surrounded by cytoplasm.
3. _____ small openings in the underside of this leaf allow the exchange of O_2 for CO_2.
4. The Haber process is _____ method for the manufacture of ammonia.
5. _____ lines on this star spectrum indicate _____ presence of different elements.
6. _____ core of the earth consists mostly of iron.
7. A photoemitter is _____ surface that releases electrons when light strikes it.
8. _____ average temperature on Io, one of Jupiter's moons, is $-148°$ C.
9. _____ patient whose lung collapsed last night should recover in a couple of months.
10. Antiseptics are _____ substances that prevent the growth of microorganisms.

The general effect of adding a relative clause or a prepositional phrase to a noun is to make the noun definite (second-mention) with *the*. If the noun in uncountable, *the* limits it to a certain location or amount. The only exception to this rule is in definitions (see the discussion of definitions in Unit IV, Section 1, "The Concrete Generic Article"), as you saw in sentences 4, 7, and 10 in Exercise 1.5.

the + NOUN	+	relative clause or prepositional phrase

Examples:
relative clauses   the experiment that was performed
                        the air that we exhale

prepositional phrases   the experiment in progress
                        the air in the mountains

EXERCISE 1.6   Add first-mention *a(n)*, *the*, or *∅* to the following sentences.

1. Microscope is optical system for observing very small objects.
2. Red spot on Jupiter may be a 13,000-mile-wide storm.
3. Symptoms that the cancer patient first noticed were tiredness and stomach pains.
4. Man who discovered oxygen was Joseph Priestley.
5. Solid is substance with a definite shape.
6. Powder on the freeway turned out to be calcium carbonate.
7. Layer below the earth's crust is called the mantle.

8. Thermometer in this system measures degrees Kelvin.
9. Boilers are devices for changing water into steam.
10. Enzyme whose structure the researchers hoped to determine was a lysozyme.

Articles With *of-Phrases*

Try the following exercise before we discuss the effect of *of*-phrases on article choice.

EXERCISE 1.7   Add first-mention *a(n)*, *the*, or *∅* to the blanks.

1. _____ atmosphere of Pluto may contain methane and argon.
2. _____ maximum depth of the Caspian Sea is 946 meters.
3. _____ pound of water weighs 454 grams.
4. _____ roots of the eucalyptus trees were not affected by last year's severe temperatures.
5. _____ piece of filter paper fell into the flask.
6. _____ height of the World Trade Center in New York is 419 meters.
7. _____ molecules of NaCl can form a lattice structure.
8. _____ center of the Milky Way galaxy is located in _____ constellation Sagittarius.
9. _____ rotation of crops protects the soil in temperate climates.
10. _____ carboxypeptidase A is _____ single polypeptide chain of 307 amino-acid residues.

An *of*-phrase is a prepositional phrase that modifies a head noun. In the following example, the head noun *circumference* is modified by the *of*-phrase *of a circle:*

HEAD NOUN   MODIFYING NOUN
the circumference of a circle

There are two major types of *of*-phrases: descriptive *of*-phrases and partitive *of*-phrases.

Articles with descriptive *of*-phrases. If the head noun is an abstract noun (for example, *wiring, study, limitation*), the *of*-phrase is almost always descriptive and must be preceded by *the*.

EXERCISE 1.8   Add the correct first-mention article to the following phrases. Underline the abstract noun in the phrase.

1. _____ combination of the lock
2. _____ precipitation of rain
3. _____ mystery of the Loch Ness monster
4. _____ discovery of radioactivity

5. _____ function of the pineal gland
6. _____ manufacture of cooling towers
7. _____ development of root hairs
8. _____ immunization of young children
9. _____ electron magnification of a protein molecule
10. _____ design of earth dams

Articles with partitive of-phrases. If the head noun is a container, or a word representing a piece of or a part of something (hence the word *partitive*), the *of*-phrase is not descriptive. Instead, it represents the noun that is divided or contained, and it requires the indefinite article *a(n)* or *∅*.

Examples:
container   a glass of water
part   a molecule of sulfuric acid

EXERCISE 1.9   Add the correct first-mention article to the following phrases. Underline the partitive head noun in the phrase.

1. _____ beaker of nitric acid (HNO_3)
2. _____ ray of sunlight
3. _____ pieces of cobalt
4. _____ box of tenpenny nails
5. _____ splinter of glass
6. _____ vial of blood serum
7. _____ isotopes of uranium
8. _____ cloud of cosmic-dust particles
9. _____ solution of NaOH
10. _____ tanks of liquid natural gas

Descriptive *of*-phrases are similar to those nouns that follow second-mention rules the first time they appear (see Unit II, Section 1, "Ranking Adjectives"). Partitive *of*-phrases are similar to first-mention nouns. When they are mentioned for the second time, they require *the*.

The company ordered a tank of oxygen. However, when the delivery truck arrived, the tank of oxygen was empty.

If the head noun is a measurement noun (see Unit II, Section 5, "Dimension Statements and Measurement Words"), the *of*-phrase acts like a partitive phrase when it indicates an amount. This is because an amount represents a limitation or part.

A resistance of twenty-eight ohms would solve the problem.
Pressures of 5700 psi are common at the bottom of the ocean.

However, some measurement nouns are not always associated with an amount. Such measurement nouns include the following words:

approximation	fluctuation*	precision	symbol
constant	increase*	reading	total
decrease*	increment	result	trace
estimate	lack	scale	variability*
excess	measurement	sign	variation*

> The lab found <u>a trace of blood</u> in the patient's urine.
> The color of a star gives <u>an approximation of its temperature</u>.

The starred (*) nouns above commonly occur with the preposition *in* when an amount is not indicated:

> The investigators found *an increase of 25 percent.*
> The investigators found *an increase in pressure.*

EXERCISE 1.10   Add *a(n), the,* or *Ø* to the blanks.

1. A U.S. household socket has _____ voltage of 110–117 volts AC.
2. _____ gravity is the force that keeps us on our planet.
3. Mercury has _____ relative density of 13.6.
4. _____ air distance between New York and San Francisco is 2571 miles.
5. _____ pressure that a woman's high heel can exert is sometimes close to 10,000 pounds per square inch.
6. A long-distance runner must have _____ strength and _____ endurance.
7. _____ excess of cholesterol in the diet can lead to blocked arteries.
8. This rope has _____ tensile strength of 5000 pounds.
9. Under the right conditions, bamboo can attain _____ growth rate of one inch per day.
10. _____ electric current (I) is determined by dividing the potential difference (V) by the impedance (Z).
11. _____ increase in global temperature would melt the polar ice caps.

Of-Phrases and the possessive/nonpossessive test.   A simple test can help you to determine if the head noun is descriptive (requiring *the*) or partitive (requiring *a(n)* or *Ø*). This test can be applied only to nouns that are **not** abstract (including gerunds, excluding measurement nouns).

It is possible to invert *of*-phrases into a form in which the head noun **follows** the object of the preposition *of*:

Descriptive: the circumference ⟶ a circle's circumference
               of a circle   (possessive)

Partitive: a molecule of ⟶ a sulfuric-acid molecule
               sulfuric acid   (nonpossessive)

Notice that the head noun following the descriptive *of*-phrase becomes a possessive noun phrase, but the head noun following the partitive *of*-phrase becomes a nonpossessive noun phrase (see Unit VI, Section 3, "Noun Compounds").

You can apply this inversion test to determine the correct article before a head noun followed by an *of*-phrase. Look at these examples:

1. _____ poles of the planet
2. _____ length of eight feet
3. _____ mountains of granite

We apply the possessive/nonpossessive test by mentally inverting the word order. In the first example, we ask ourselves if it is possible to say "the planet's poles." Is this a logical (although possibly uncommon) possessive? Do the poles "belong" to the planet? The answer is yes. The poles belong to the planet as an attribute. Therefore, the article must be *the:*

1. __the__ poles of the planet

In the second example, is it logical to say "an eight foot's length"? Clearly, the answer is no; we would say "an eight-foot length," the nonpossessive form. Therefore, the answer must be *a(n):*

2. __a__ length of eight feet

In the third example, is it logical to say "granite's mountains"? Again, the mountain does not "belong" to the granite. We would say "granite mountains," the nonpossessive form. Therefore, the answer must be *∅:*

3. __∅__ mountains of granite

The possessive/nonpossessive test is accurate in predicting the correct article before most head nouns with *of*-phrases. Remember, however, that if the head noun is an abstract noun (except measurement nouns) or a gerund, the article must be *the* and the possessive/nonpossessive test is not required. Remember also that the inversion may not be a commonly written form. For example,

The probe examined the surface of the third largest nonferrous asteroid.

We would definitely not use the phrase "the third largest nonferrous asteroid's surface." However, the phrase "the asteroid's surface" is logically possessive, thereby making the correct article *the*. It is the logic of the relationship, not the acceptability of the written form, that is important.

EXERCISE 1.11 Assuming that the following phrases are being mentioned for the first time, put the appropriate article in the blank.

1. _____ area of a circle
2. _____ detailed photograph of the rings of Saturn
3. _____ poles of the planet
4. _____ hundreds of tentative hypotheses
5. _____ lack of large impact craters
6. _____ no signs of turbulence
7. _____ fluctuation of 10 percent
8. _____ most dramatic result of iron deficiency
9. _____ mixture of several solvents
10. _____ cancer of the mouth
11. _____ significance of the study
12. _____ velocity of 100 miles per hour
13. _____ its rate of growth
14. _____ purpose of an experiment
15. _____ plane of our galaxy
16. _____ name of the company
17. _____ reverse side of this page
18. _____ new estimates of the damage
19. _____ three-year supplement of iodine
20. _____ site of the space-shuttle landing

EXERCISE 1.12 Fill the blanks with *a(n)*, *the*, or *0*.

THE PLANET EARTH

The earth is made almost entirely of rock and metal. _____ outside of the earth has _____ thin covering of soil. Inside this covering, there is _____ layer of solid rock 30 to 50 miles deep. Below this crust of soil and rock lies _____ mantle of the earth, which has _____ thickness of about 600 miles. Next comes an intermediate layer with _____ depth of about 1200 miles. Scientists have learned something about this layer by tracing _____ patterns of earthquake shocks. At _____ center of the earth lies the white-hot core. It has _____ pressure of 45 million pounds per square inch and consists mostly of iron. Scientists believe that the earth is made of the same material as _____ smaller members of the solar system called meteorites. All these solid and liquid layers are held together by _____ force of gravity.

Section 2
SENTENCE COMBINING

NONDEFINING RELATIVE CLAUSES

In Units I and II, we looked exclusively at defining relative clauses. A defining relative clause, like a definition, identifies a noun as being different from other nouns. For example, the sentence

> The beaker that contains nitric acid is on the left.

gives us two pieces of information: (1) the beaker contains nitric acid, and (2) there must be more than one beaker present—otherwise there would be no need to define it (we would just say, "The beaker is on the left."). We imagine, therefore, that the speaker is identifying the beaker in order to warn us. Perhaps there is also a beaker containing water or another colorless liquid, and the speaker does not want us to confuse them.

A nondefining relative clause is distinguished from a defining one by the use of commas. It does not limit a noun, like a defining relative clause does. It simply gives us additional information about that noun. For example, the sentence

> The element plutonium, which was first produced at U.C. Berkeley, is a potent carcinogen.

also gives us two pieces of information: (1) plutonium is a carcinogen that was first produced at U.C. Berkeley, and (2) there is only one element called plutonium. If we removed the commas, the resulting defining relative clause would indicate that there is one kind of plutonium that was produced at U.C. Berkeley and imply that there is another kind produced somewhere else. That would be incorrect. Another example:

> The beaker, which contains nitric acid, should be handled with care.

In this case, although there could be other beakers in the lab, the speaker/writer tells us by using commas that he or she is concerned only with this one beaker. We have no reason to imagine the presence of any other beakers.

A nondefining relative clause is always used if the head noun is a proper noun (see Unit VI, Section 1):

> The Lightman Chemical Company, which is located in New Jersey, produces high-quality reagents.

> Professor White, who is well known for his medical research, gave a lecture at the conference last week.

This is because a proper noun indicates one individual entity and no other.

EXERCISE 2.1   Underline the relative clause in the following sentences. Then label the sentence defining (D), or nondefining (ND).

Example: ___   The planet that they observed was Venus.
Answer: <u>D</u>   The planet <u>that they observed</u> was Venus.

1. _____ The minerals that the human body requires are usually obtained from plants.
2. _____ The Lawrence Berkeley Laboratory, which is located near the University of California, was established in 1936.
3. _____ The narrow band around the earth in which life can exist is called the biosphere.
4. _____ The oxygen tank that was delivered yesterday has a leaky valve.
5. _____ Hydrofluoric acid, which is made up of hydrogen and fluorine, is used chiefly to etch glass.
6. _____ The chemicals that make soft sea creatures unpalatable to predators are of interest to scientists.
7. _____ Transistors, which are really tiny amplifiers, use very little power.
8. _____ Surgical blades that are made from volcanic glass are sharper than diamond scalpels.
9. _____ People who smoke have reduced life expectancies.
10. _____ Vega, which is the brightest star in the constellation Lyra, may be generating a solar system.

EXERCISE 2.2   Add commas to the following relative clauses where necessary. Describe the implication (Imp:) of the sentence.

Example: The beaker that contains sulfuric acid should be covered.
Answer: The beaker that contains sulfuric acid should be covered.
Implication: Other beakers containing other substances are present.

1. Patients who have infectious diseases are usually isolated.
   Imp:
2. The energy which is contained in wood can be released by burning.
   Imp:
3. Bauxite which is a claylike material is an ore that contains aluminum.
   Imp:
4. The element which has the highest melting point is tungsten (W).
   Imp:
5. Mechanical springs which are used in machines can be classified as either wire springs or flat springs.
   Imp:

6. Osmium which has the highest specific gravity of any element was discovered in 1803.
   Imp:
7. An illness which is caused by bacteria can usually be cured by means of antibiotics.
   Imp:
8. The first atomic bomb which equaled 20,000 tons of TNT melted sand into glass up to 800 yards away.
   Imp:
9. A V-belt which is usually made of fabric and rubber may be used with small pulleys.
   Imp:
10. A V-belt which is worn should be replaced immediately.
    Imp:

Subject-Form and Object-Form Nondefining Relative Clauses

Like the defining relative clauses, nondefining relative clauses have both subject and object forms:

S-form: The tomato plant, which requires a lot of sunlight, was first grown by Indians in the Andes.
O-form: Nuclear magnetic resonance, which doctors use to obtain images of internal organs, is a relatively recent invention.

Nondefining relative clauses differ from defining relative clauses in that the relative pronoun *that* may not be used.

NATURE OF NOUN	RELATIVE PRONOUNS	
	Defining	Nondefining
Person	who	who
Person	whom	whom
Thing	which	which
Person/Thing	that	---
Person/Thing	whose	whose
Place	where	where
Time	when	when

In referring to things, some scientific writers prefer to use the relative pronoun *that* for defining relative clauses and the pronoun *which* for nondefining ones. This is a useful and convenient distinction, but unfortunately it is not universally used. You will therefore often see *which* used in both defining and nondefining relative clauses. However, you will never see *that* used with a nondefining relative clause.

EXERCISE 2.3   Combine the following sentences into nondefining relative clauses.

Example: Water is required by all living things. It is composed of hydrogen and oxygen.
  Answer: Water, which is composed of hydrogen and oxygen, is required by all living things.

1. Albert Einstein discovered the equivalence of mass and energy. He was born in Germany in 1879.
2. The Pacific "ring of fire" extends from California to Japan. Most earthquakes take place there.
3. Adrenaline was first isolated by Jokichi Takamine. Its secretion is stimulated by strong emotion.
4. Wyandotte Cave contains one of the world's largest underground mountains. It is located in southern Indiana in the U.S.
5. Gasohol is a mixture of gasoline and alcohol. Even the Ford Model T could run on it.
6. The Paleozoic era lasted 355 million years. Creatures left the sea then.
7. The Arecibo radio telescope is situated in Puerto Rico. It has a diameter of 1001 feet.
8. Iridium is 1000 times more abundant in meteorites. Scientists have found it between rocks of the Cretaceous and Tertiary periods.
9. The U.S death rate from heart disease was 330.4 per 100,000 in 1979. It was 153.0 per 100,000 in 1900.
10. A force of attraction exists in the universe between each body and every other body. We call this force gravity.

Section 3
REDUCTION AND MODIFICATION

REDUCING NONDEFINING RELATIVE CLAUSES

Nondefining relative clauses cannot always be reduced in the same way as defining relative clauses. Let us compare the rules for reducing defining relative clauses (Unit II, Section 3) with the rules for reducing similar nondefining relative clauses in (1) object form and (2) subject form.

Reducing Object-Form Nondefining Relative Clauses

Defining Rule: Remove relative pronoun in an O-form relative clause.
Examples:
Defining: The report the lab first sent us contained serious errors.

Nondefining:
> Incorrect: *The report, *Scientific American* <u>published last month</u>, contained several errors.
> Correct: The report, <u>which</u> *Scientific American* <u>published last month</u>, contained several errors.

Nondefining Rule: Object-form nondefining relative clauses may not
[different]   be reduced.

EXERCISE 3.1   Reduce the following defining and nondefining relative clauses where possible.

Example: The food that we eat should be varied.
 Answer: The food we eat should be varied.

1. The mouse that the pharmacologist inoculated did not develop leukemia.
2. Hurricane Alicia, which meteorologists warned the public about in advance, did extensive damage.
3. The Landsat 4 satellite, which NASA launched in 1982, might fall back to earth.
4. The DNA that researchers have analyzed in mitochondria is used in studies of human evolution.
5. Halide glasses, which optical-fiber experts expect to replace glass fibers, are formed from combinations of halogens and metals such as zirconium and hafnium.
6. Cigarettes that the tobacco industry claims are low in nicotine are not safer than other cigarettes.

Reducing Subject-Form Nondefining Relative Clauses

Subject-form relative clauses can be reduced in two ways: (1) by removing the relative pronoun + *be* under certain conditions and (2) by removing the relative pronoun and changing the verb to the V_{ing} form.

Removing the relative pronoun + *be.*

> Condition 1: Remove relative pronoun + *be* if the words following *be* consist of a past participle (V_{ed_2}) with a modifying phrase (i.e., a passive relative clause).
> Examples:
>> Defining: The plutonium <u>stolen from the lab</u> was never found.
>> Nondefining: The new shipment of plutonium, <u>stolen from the CERN lab</u>, has still not been found.
>
> Nondefining Rule: Remove relative pronoun + *be* if the words follow-
> **[same]**   ing *be* consist of a past participle (V_{ed_2}) and a modifying phrase.

EXERCISE 3.2 Reduce the following defining and nondefining relative clauses.

1. Saturn's moon Hyperion, which has been forced to tumble by tidal friction, is shaped like an oval hamburger.
2. Unburned hydrocarbons that are released by automobiles contribute to the production of smog.
3. Hemoglobin, which is made up of four peptide chains, carries oxygen through the body.
4. Diseases that occur in crops can be controlled by chemicals and good farm practices.
5. Wind tunnels, which are used to check metal fatigue in planes, can develop air speeds that are many times the speed of sound.
6. Some recently discovered tools and animal bones, which were deposited in East Africa approximately two million years ago, might indicate the beginning of human social organization.

Condition 2: Remove relative pronoun + *be* if the word following *be* is one of a group of past participles that show a specific or a temporary state or refer to something mentioned earlier (e.g., *shown, produced, given, used*).

Examples:
    Defining: The graph <u>shown</u> represents the final results.
Nondefining:
    Incorrect: *The graph, <u>shown</u>, represents the final results.
    Correct: The graph, <u>shown on page 24</u>, represents the final results.

Nondefining Rule: Remove relative pronoun + *be* **only** if *be* is followed
[**different**]   by one of a group of past participles showing a specific or a temporary state or referring to something mentioned earlier **plus a postmodifying phrase**.

EXERCISE 3.3 Reduce the following sentences, filling in the blanks with a phrase **if necessary**.

Example: One cell, which is shown ——————————— , is beginning to divide.
Answer: One cell, shown in the photograph, is beginning to divide.

1. The diagram, which is explained ——————————— , shows the four phases of an internal-combustion engine.
2. The words that have been underlined ——————————— all contain errors.
3. The islands, which were discovered ——————————— , were all of volcanic origin.
4. The agent that was found ——————————— was a variant of the swine-flu virus.

5. The refining procedure, which was described _____ , involved several distillations.
6. The herbicide that is used _____ does not contain DDT.

Condition 3: Remove relative pronoun + *be* if the words following *be* consist of a present participle (V_{ing}) with a postmodifying phrase.
Examples:
> Defining: The tropical storm <u>approaching the coast</u> will cause severe flooding.
> Nondefining: Hurricane Robert, <u>approaching the coast</u>, will cause severe flooding.

Nondefining Rule: Remove relative pronoun + *be* if the words following *be* consist of a present participle (V_{ing}) with a
[**same**] postmodifying phrase.

EXERCISE 3.4   Reduce the following defining and nondefining relative clauses.

1. The doctor, who is working with disturbed adult patients, has discovered a link between early-life stress and major psychiatric illness.
2. The substance that is blocking the artery is largely cholesterol.
3. The tanker, which was carrying a million gallons of crude oil, broke up in heavy seas off the coast of France.
4. The planes that are spraying malathion are being used to control the Mediterranean fruit fly.
5. The city transit system, which was operating at a deficit, could not afford to extend its service.
6. The tumor, which is lying directly against the liver, cannot be removed without serious damage.

Condition 4: Remove relative pronoun + *be* if the words following *be* constitute a prepositional phrase.
Examples:
> Defining: A body <u>at rest</u> has no motion in relation to an observer.
Nondefining:
> Incorrect: *The pendulum, <u>at rest</u>, is being recalibrated.
> Correct: The pendulum, <u>temporarily at rest</u>, is being recalibrated.

Nondefining Rule: Remove relative pronoun + *be* **only** if the words following *be* consist of **an adverb** followed by a prep-
[**different**] ositional phrase.

EXERCISE 3.5   Reduce the following defining and nondefining relative clauses.

1. A new aerosol vaccine that is against measles is highly effective in six-to-twelve-month-old infants.
2. Pluto, which is normally at the outer edge of the solar system, is a very cold planet.
3. The scratch, which is on the surface, should not affect your vision.
4. Aluminum, which is close to magnesium in atomic weight, has a much higher melting point.
5. The difference that is between a star and a planet depends mainly on whether the body has enough mass to initiate nuclear burning.
6. The xylem, which is usually situated outside the phloem, carries water and minerals from the roots to the leaves.

Condition 5: Remove relative pronoun + *be* if the word following *be* belongs to a special group of adjectives ending with *-ble* (e.g., *possible, responsible, visible, capable*).

Examples:
Defining: A small scar was the only change <u>visible</u> three weeks after the operation.
Nondefining:
Incorrect: *Dr. Smith, <u>responsible</u>, is unavailable for comment.
Correct: Dr. Smith, <u>responsible for the cancer ward</u>, was unavailable for comment.

Nondefining Rule: Remove relative pronoun + *be* **only** if the word following *be* is one of a special group of adjectives ending in *-ble* **plus a modifying phrase**.

[**different**]

EXERCISE 3.6   Reduce the following defining and nondefining relative clauses.

1. The space telescope, which is capable of seeing farther than any land-based observatory, will receive funding for only ten years.
2. The virus that is responsible may be related to the measles virus.
3. Hieroglyphics, which were unreadable before the discovery of the Rosetta stone, revealed many aspects of ancient Egyptian life.
4. The IRAS-Araki-Alcock comet, which was visible for a few days in May 1983, passed within 2.9 million miles of earth.
5. The only explanation that was possible was that there had been an error in the procedure.
6. The lens of the eye, which is remarkable for its transparency, consists of closely packed proteins.

Removing the relative pronoun and changing the verb to V_{ing}.

Condition: Remove relative pronoun and change the verb to the V_{ing} form as long as the statement represents a fact (not just a single event) and the verb is **not** *be* (see Unit V, Section 2, "Dangling Modifiers").

Examples:

> Defining: The planets <u>circling the sun</u> constitute the solar system.
>
> Nondefining: Titan, <u>circling the planet Saturn</u>, may have an ocean of liquid nitrogen.
>
> Nondefining Rule: Remove relative pronoun and change the
> [**same**] verb to the V_{ing} form as long as the statement represents a fact (not just a single event) and the verb is **not** *be*.

EXERCISE 3.7   Reduce the following defining and nondefining relative clauses.

Example: The man who teaches physics is from the Soviet Union.
Answer: The man teaching physics is from the Soviet Union.

1. Soft drinks that contain phosphates (e.g., cola) can prevent lead (Pb) poisoning.
2. Vent microbes, which grow near sulfide chimneys on the ocean floor, can thrive at temperataures of 250° C or higher.
3. A chemical that mimics an insect pheromone is being used to control certain agricultural pests.
4. The Eustachian tubes, which connect the ears to the throat, equalize the pressure on the eardrum.
5. The acid rain that occurs in eastern Canada and the U.S. has harmed many glacial lakes.
6. The orbit of a binary-star system, which decays with age due to tidal interactions, may produce a planetary nebula.
7. If the earth did not rotate, the side of the planet that faces the sun would be the only source of winds.
8. The first commercial engine that operated on steam was based on the principles of Papin, a French physicist.
9. An unmagnetized ferromagnetic substance has magnetic axes that point in various directions.
10. A retinal rod in the human eye contains 10^9 rhodopsin molecules, which guarantee that even one photon will elicit a nerve impulse.

EXERCISE 3.8   Make all possible relative-clause reductions in the following passage.

CORAL REEFS[2]

Coral grows in all seas, but the reef-building varieties, which are never deeper than 150 feet, which are rarely at temperatures that are below 68° F., and which are seldom more than 1500 miles from the equator, are only found in tropical waters. Corals must have sunlight for the algae which grow in their

---

[2]Adapted from Jane Werner Watson, *The World We Live In* (London: Wm. Collins Sons, 1957), pp. 116–17.

tissues. They must also have clean water because sediment and mud suffocate them. Moving water is needed to bring them sufficient food and oxygen. They are flesh eaters, which feed on young fish, tiny shelled creatures, and sea worms, which they catch with stinging tentacles.

Three typical creations of coral are fringing reefs, barrier reefs, and atolls. Fringing reefs, which fan out from the edge of land in an almost solid shelf, typically surround old volcanic islands. Barrier reefs, which are separated from land by a wide lagoon or channel, are best represented by the Great Barrier Reef in Australia. Atolls, which are built out from sunken islands, are coral rings which enclose lagoons in the open sea. The rock which is under some atolls extends 4000 feet down into the sea.

Great coral reefs are found not only in modern seas but in many land areas that were under water millions of years ago. A great series of coral reefs, which were formed in the Paleozoic era, has been discovered in western Canada.

MOST, MOSTLY, ALMOST, ETC.

The modifiers based on the word *most* can be divided into two major groups: (1) those that indicate "the greater part," "a clear majority," or 70 to 90 percent and (2) those that indicate "superlative degree" (see Unit VI, Section 3, "Comparisons") or 100 percent. These modifiers are of two types: determiners (det) or adverbs (adv).

Majority or 70 to 90 Percent

The following words and phrases indicate 70 to 90 percent:

1. *most* (det)   <u>Most</u> animals have nervous systems. (general)
2. *almost all* (det)   <u>Almost all</u> plants contain chlorophyll. (general)
3. *most of* (det)   <u>Most of</u> the trees in this region are deciduous. (specific)
4. *almost all (of)* (det)   <u>Almost all (of)</u> the cells in this sample are cancerous. (specific)
5. *mostly* (adv)   Bronze consists <u>mostly</u> of copper (Cu).
6. *almost* (adv)   The dam is <u>almost</u> finished [i.e., not quite finished].
7. *for the most part* (adv)   The water on the earth is <u>for the most part</u> salt water.

Superlative Degree or 100 Percent

8. *the most* (adv)   Mexico City is <u>the most</u> populated city in the world.
9. *most* (adv)   The medicine doctors prescribe <u>most</u> is aspirin.

10. *most of all*   Giant pandas eat various plants but prefer bamboo flowers <u>most</u>
    (adv)   <u>of all</u>.
11. *at most*   The gestation period of a rat is <u>at most</u> twenty-one days [i.e.,
    (adv)   twenty-one days is the maximum].

Notice that the determiners modify a noun and the adverbs modify a verb or an adjective.

EXERCISE 3.9   For each of the following sentences, choose all the possible modifiers from the list below:

almost	for the most part	mostly
almost all (of)	most	most of all
at most	most of	the most

Example: _____ this oil is from Egypt.
Answer: Most of/Almost all of/For the most part this oil is from Egypt.

1. _____ the elements with an atomic number between 50 and 70 are rare-earth elements.
2. The California condor had _____ disappeared before a new captive-breeding program was established.
3. In medicine, nuclear magnetic resonance is used _____ for creating images of internal organs.
4. Recombinant-DNA techniques have produced interferon and other substances, but the product that has been manufactured _____ in this way is insulin.
5. _____ the arthropods in the late Silurian period were spiders, scorpions, and millipedes.
6. Acid rain occurs _____ in the eastern two thirds of the North American continent.
7. The nuclear reactor at Three-Mile Island _____ had a meltdown.
8. The data currently in the computer memory are[3] _____ digitized images from Saturn.
9. _____ good engineers enjoy their work.
10. The geographical area with _____ rainfall per year is the Amazon Basin.

[3]The Latin word *data* is a plural form. The singular form is *datum*. (These data are correct. This datum is correct.) Many writers avoid the word *datum* and use only the plural form *data*. However, *data* is used by many scientists as a mass noun, which is a singular form. (The data is correct.) This usage is not universal; it is in the process of change. It is, therefore, safer in formal writing to use *data* as a plural countable noun, although you will often hear it used as a mass noun.

11. The results of the experiment corresponded _____ exactly with the results found by the Swedish team.
12. Air pollution in the form of hydrocarbons comes in part from road vehicles and electrical utilities, but it comes from industrial processes _____ .
13. Human beings can live without oxygen for _____ six minutes without sustaining brain damage.

EXERCISE 3.10   Transform the sentences from Exercise 3.9 into the sentences below.

Example: Most of this oil is from Egypt.
            This oil _____ .
Answer: This oil is mostly from Egypt.

1. The elements _____
   _____ .
2. _____ California condors were extinct _____
   _____ .
3. _____ frequent application of nuclear magnetic resonance is for _____ .
4. [No change possible.]
5. The arthropods in the late Silurian period _____
   _____ .
6. _____ acid rain _____
   _____ .
7. [No change possible.]
8. _____ the data _____
   _____ .
9. Good engineers _____ .
10. It rains _____ in the Amazon Basin.
11. [No change possible.]
12. The source of air pollution in the form of hydrocarbons is _____
    _____ .
13. [No change possible.]

PREPOSITIONS: *at, on,* AND *in*

We will begin our study of the prepositions by concentrating on three very common ones: *at, on,* and *in.*

At

The preposition *at* commonly refers to a position or to a location that implies a function.

Position of time, place, or measure.

1. The rocket was launched <u>at 5:39 A.M.</u>
2. The neutron is <u>at the center</u> of the atom.
3. Sulfur melts <u>at a temperature</u> of 113° C. (*At* plus a measurement noun is a common structure.)

Other Uses

at the age of	*at* a level	*at* the surface
at the beginning	*at* a node	*at* the terminal
at the bottom	*at* a point	*at* the top
at the end	*at* a stage	

Location that implies function

Examples

1. The professor is <u>at the lab</u> today. (The professor may be inside or outside the actual lab. This sentence indicates that the professor is doing the things one normally does in a lab.)
2. The assistant is <u>at the post office</u> (The assistant is doing what one normally does at a post office, such as buying stamps, mailing a letter, or picking up a package.)

Other uses

at the airport	*at* the garage	*at* school
at the bank	*at* home	*at* the store
at court	*at* the hospital	*at* work

In relation to time, the following diagram may help you to remember the uses of *at, on,* and *in:*

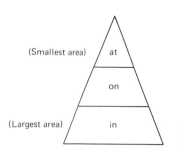

(Smallest area)   at

on

(Largest area)   in

time: The rocket was launched <u>at 5:39</u>.

day: The meeting is <u>on Tuesday</u>.
date: The first atomic explosion took place <u>on July 16, 1945</u>.

month: Sunspots are most active <u>in May</u>.
season: It usually snows here <u>in winter</u>.
year: The cyclotron was built <u>in 1931</u>.
decade: Computers become miniaturized <u>in the 1970s</u>.
century: Sir Isaac Newton was born <u>in the seventeenth century (the 1600s)</u>.

On

The preposition *on* commonly refers to a surface or a line:

Examples
1. Check the reading <u>on the barometer</u>.
2. Oxides <u>on the surface</u> prevent the metal from corroding.
3. Building <u>on a fault line</u> is not advised.

Other Uses		
on the area	*on* the gauge	*on* a plane
on the face (of)	*on* a line	*on* a scale
on the edge	*on* a path	*on* a side
on the verge (of)	*on* the average	

In

The preposition *in* commonly refers to containment or mode.

Containment.

Examples
1. The sample is <u>in the test tube</u>.
2. Ozone occurs <u>in the upper atmosphere</u> (a limited area).
3. The program failed because of bugs <u>in the system</u>.

Other Uses		
in the body	*in* a manometer	*in* a position
in this case	*in* the ocean	*in* a range
in the circuit	*in* a plane	*in* a situation
in a liquid	*in* this problem	*in* a tank

Mode: measurement, direction, action.

Examples
1. The box is two feet <u>in length</u>. (See Unit II, Section 5, "Dimension Statements and Measurement Words.")
2. The blood is constantly <u>in motion</u>.

3. The error occurred <u>in preparing</u> the data. ($In + V_{ing}$ is a common form.)

	Other Uses	
in boldface type	*in* a form	*in* proportion
in a direction	*in* a manner	*in* question
in equilibrium	*in* operation	*in* series
in fact	*in* parallel	*in* a way

EXERCISE 3.11 Add *at, on,* or *in* to the blanks.

1. Is there a black hole _____ the center of the Milky Way?
2. The great San Francisco earthquake occurred _____ April 18, 1906.
3. The Boeing 747 is 231.9 feet _____ length.
4. The pressure _____ the bottom of the deepest part of the Pacific Ocean is 16,380 pounds per square inch.
5. The planets _____ the solar system lie for the most part ____ a plane.
6. Some researchers claim to be _____ the verge of discovering the cause of some forms of cancer.
7. The depth of the oceans is 2.5 miles _____ the average.
8. Large-scale cooling towers are usually constructed _____ the site.
9. The flow of a fluid _____ a pipeline can be measured with a venturi meter.
10. Enzymes bind to substrates _____ very specific locations.
11. Resistors connected _____ parallel have a much lower overall resistance than those connected _____ series.
12. A force acting _____ a body produces an acceleration _____ the direction of the force.
13. The reading _____ a barometer indicates atmospheric pressure.
14. Hydrogen boils _____ a temperature of $-252.87°$ C.
15. The gases _____ the atmosphere are primarily nitrogen and oxygen.
16. The specimen was mounted _____ a glass slide for microscopic analysis.
17. _____ 100,000 feet, 99 percent of the mass of the atmosphere is below.
18. A human embryo is considered to be full term _____ forty weeks.
19. A body _____ motion has a momentum equal to the product of its mass and its velocity.
20. A force is a uniform motion _____ a straight line.
21. A projectile moves _____ a parabolic path that falls _____ the midpoint because of air resistance and gravity.

22. The meniscus is located _____ the surface of a fluid _____ a tube.
23. Craters _____ the surface of the moon were probably caused by falling meteorites.
24. Monkeys typically die _____ an age of 13 years.
25. Hibernation in the Northern Hemisphere usually ends _____ March.

Section 4
VERBS

PAST TENSES

The most common past-tense form in scientific English is the simple past tense.

The Simple Past Tense

In scientific English, the simple past tense is commonly used to describe past research. Thus, it is often used in the background section of the introduction to a research report. It is also used to describe past discoveries, and because discoveries often represent descriptions of a process, the simple past tense is introduced in this unit.

The simple-past verb forms (V_{ed_1}) are either regular or irregular. The regular form ends with -ed. The irregular form has many different spellings (see Appendix A). However, the singular and plural forms of both regular and irregular verbs are identical.

EXERCISE 4.1 Fill the blanks with the correct simple-past form of the verbs in parentheses.

THE WRIGHT BROTHERS[4]

The Wright brothers (undertake) _____ to design and build a machine that (will fly) _____ _____ . They (design) _____ the entire aircraft, including the power plant and the propeller. Basic research, applied research, development—all of these they (carry out) _____ _____ themselves. They (find) _____ that they (cannot) _____ _____ safely accept what others (say) _____ until they themselves (verify) _____ it. They (have) _____

[4]Walter J. Miller and Leo E. A. Saidler, *Engineers as Writers* (New York: Van Nostrand, 1953), pp. 302–3.

meagre financial resources and little formal engineering education. Yet, the Wrights (build) _____ and (fly) _____ their airplane successfully because they (see) _____ the problem as a whole, (give) _____ it the attention it (deserve) _____ and (overlook) _____ nothing.

The Past Continuous and Past Perfect Tenses

In addition to the simple past tense, there are two other past tenses whose purpose is to show the relationship between two past events: (1) the past continuous tense and (2) the past perfect tense.

Past continuous tense.   The past continuous tense is used to show that two actions occurred at the same time. This tense usually occurs with a simple-past verb that establishes the time reference. The past continuous form represents a continuous process. The simple-past form represents a single action.

> The beaker broke when the water was boiling.
> The geologist was taking photographs when the volcano erupted.

Past perfect tense.   The past perfect tense is used to show that one action preceded another in the past. This tense also occurs with a simple-past verb that establishes the time reference. The past perfect form shows the first event. The simple past form shows the second event.

> The researcher reviewed the paper she had written.
> When the engineer had finished the investigation, a well-known journal published the results.
> All the glassware was on the floor. The earthquake had struck early that morning.

EXERCISE 4.2   Fill the blanks with the correct past form of the verbs in parentheses.

THE DISCOVERY OF PENICILLIN[5]

Alexander Fleming, a Scottish research bacteriologist, (study) _____ the deadly staphylococcus when he (make) _____ his famous discovery in 1928. For examination purposes, Fleming (remove) _____ the cover of the bacteria culture with which he

[5]Adapted from "Fleming," in *Compton's Pictured Encyclopedia* © 1964 by Encyclopaedia Britannica, Inc.

(work) _____ _____. A mold (form) _____
on the exposed culture. Fleming (notice) _____ that in
the area surrounding the mold, the staphylococci (disappear) _____
_____. He (keep) _____ a strain of the mold alive
and (begin) _____ testing it on laboratory animals. In
1929, he (publish) _____ his first medical paper in
which he (prove) _____ that a simple soil mold (be)
_____ a powerful microbe killer that (not injure) _____
_____ _____ human tissue.

For years, chemists (be unable) _____ _____to
extract enough pure concentrated penicillin to use in medicine. Then in 1938,
a team of Oxford scientists (remember) _____ the re-
search paper of nine years earlier. They ultimately (succeed in) _____
_____ developing a method for mass-producing penicillin.

INFINITIVE STRUCTURES

When a verb is preceded by a subject, the verb is said to be finite—that is, it is
limited by that subject. For example,

SUBJECT   FINITE VERB
Heat   rises.

If we remove the subject, the verb is no longer limited or finite but rather un-
limited or infinite. This is why we call such a verb the infinitive form.

         FINITE   **INFINITE**
Heat   rises.   →   to rise

Infinitives have two forms: (1) with *to* (the standard form) and (2) without *to*
(the bare or *to*-less form).

Infinitives With *To*

The standard form of the infinitive verb is *to*+an uninflected verb (a
verb that has not been altered, or inflected, to indicate singular/plural or past/
present). Infinitives occur as the result of sentence combining in which we remove a repeated noun phrase, leaving the verb infinite. Examples:

1. Repeated subject noun phrase:

   The doctors want X.
   + X = The doctors will buy an X-ray machine.

   _____
   X =   (delete)   to buy an X-ray machine.
   The doctors want to buy an X-ray machine.

2. Repeated object noun phrase:

The doctors want the hospital X.
+ X = The hospital will buy an X-ray machine.

X =   (delete)   to buy an X-ray machine.
The doctors want the hospital to buy an X-ray machine.

Infinitives with *to* occur in four grammatical situations: (1) after certain verbs, (2) to show purpose, (3) after objects, and (4) after certain adjectives.

After certain verbs   An infinitive verb with *to* must be used after certain verbs, most of which indicate the future in some manner. These verbs include the following:

agree	*dare*	*hope*	*plan*	*seem*
arrange	*decide*	*intend*	*prefer*	*tend*
attempt	*demand*	*learn*	*prepare*	*try*
choose	*determine*	*manage*	*promise*	*want*
claim	*expect*	*need*	*refuse*	*wish*
consent	*fail*	*offer*	*resolve*	

Examples:
   The laboratory expects to have the results soon.
   They planned to shut down the reactor if there was a leak.

To show purpose.   An infinitive verb with *to* indicates a reason or purpose in answer to the question why. The phrase *in order to* has the same function, but it is usually considered to be wordy (see Unit V, Section 5, "Wordiness"). Examples:

   Pressure is applied to increase product yield.
      (Why is pressure applied?)
   Plants need sunlight to grow.
      (Why do plants need sunlight?)

After objects.   An infinitive verb with *to* often occurs after objects (see also Unit VI, Section 2). Examples:

   The nurse asked the patient to breathe deeply.
   Pressure causes the gasoline mixture to explode.

After adjectives and participles. An infinitive verb with *to* often occurs after adjectives and participles. Examples:

> The patient was <u>anxious to hear</u> the diagnosis.
> The university is <u>required to release</u> the data.

EXERCISE 4.3 Combine the following sentences, using infinitive structures.

Example: They want X. X = They will finish the experiment today.
 Answer: They want to finish the experiment today.

1. Medical researchers hope X. X = They will find a cure for cancer by the end of this century.
2. Air or gas expands. Air or gas fills the enlarged chamber and thus cools.
3. Experimental physicists like X. X = Experimental physicists find cases where established laws do not work.
4. It is relatively easy. Somebody determines the exact melting point of a solid.
5. Dentists must allow X. X = The jaw heals before implanting anchors for a fixed bridge.
6. In the grazing areas of Australia, the number of animals per acre a farmer can expect depends on the pasture of its poorest growing season. The farmer can raise a number of animals.
7. The only way that kidney-failure victims can survive is X. X = They have their kidneys cleansed or they get a kidney transplant.
8. Two groups have successfully used solar energy. Solar energy splits water.
9. High-temperature geothermal sources are hot enough. They turn the turbines of electric generators.
10. Perpendicular magnetic recording will enable X. X = Computer operators will squeeze more information onto magnetic disks, drums, and tapes.

EXERCISE 4.4 Correct the infinitive errors in the following sentences. Each sentence has at least one error.

1. Several countries plan send a probe observing the new comet.
2. Doctors need analyzing a blood sample before to make a diagnosis.
3. Space-shuttle experiments to test a plant's ability for synthesizing carbohydrates under zero-gravity conditions.
4. Some patients agreed for try the new drug a second time.

5. Camels have been known going without water for two weeks.
6. Mice learn very quickly choosing which passage leads to food.
7. Formal logic is a very difficult subject for learning a child.
8. The psychiatrist refused discussing her patient with the reporter.
9. The protons were forced accelerate to 130 MeV (million electron volts).
10. Glassware must to be cleaned after every experiment.

Infinitives Without *to* (Bare or *to*-less Infinitives)

Infinitive verb forms without *to* also occur as the result of sentence combining in which we remove a repeated object noun phrase, leaving the verb infinite. Example:

> They heard the tree X.
> +X=The tree fell.
>
> X=   (delete) fall
> They heard the tree fall.

Infinitive verb forms without *to* occur in three grammatical situations: (1) after modals (e.g., *will, should, must, may*), (2) with the causative verbs *let, make,* and *have,* and (3) after perception verbs (e.g., *see, hear, watch, feel*).

After modals. An infinitive verb without *to* must be used after a modal auxiliary such as *will, can, must, should, might,* or *would* (see also Unit VI, Section 4, "The Subjunctive"). Examples:

> The three rockets <u>must ignite</u> at precisely the same moment.
> The probe <u>will leave</u> the solar system in 1989.

With the causative verbs *let, make,* and *have*. An infinitive verb without *to* is used in the complement of the causative verbs *let, make,* and *have.* (The word *help* can function as a causative, but it also occurs with a normal infinitive).
1. *Let*=allow, to make it possible for something to happen. Examples:

> The doctor <u>let</u> the students <u>look</u> into the microscope.
> The doctor <u>allowed</u> the students <u>to look</u> into the microscope.
> The engineers <u>let</u> the gas in the reactor <u>escape</u>.
> The engineers <u>allowed</u> the gas in the reactor <u>to escape</u>.

2. *Make*=to force, to cause, to act upon with considerable pressure. Examples:

The government <u>made</u> the utility company <u>pay</u> for the accident.
The government <u>forced</u> the utility company <u>to pay</u> for the accident.
The gravity tides of Saturn and its moons <u>are making</u> some of the rings <u>twist</u> into a braided pattern.
The gravity tides of Saturn and its moons <u>are causing</u> some of the rings <u>to twist</u> into a braided pattern.

3. *Have* = to agree to do something because of an accepted condition (e.g., authority, money, personal relationship). Examples:

The professor <u>had</u> the students <u>solve</u> the problem by themselves. (The students agreed because they accepted the authority of the professor).
The professor <u>asked</u> the students <u>to solve</u> the problem by themselves.
The doctor <u>had</u> the nurse <u>give</u> the patient a sedative. (The nurse agreed because she is paid to do so and because she accepts the authority of the doctor.)
The doctor <u>asked</u> the nurse <u>to give</u> the patient a sedative.

With perception verbs. An infinitive without *to* is used in the complement of the perception verbs *see, watch, notice, hear, feel,* and *observe* (with V_{ing}). Examples:

The physicists <u>saw</u> the particle <u>explode</u> as it hit the neutron.
The biologists <u>watched</u> the cell <u>divide</u> under the microscope.
The geologists <u>heard</u> the rock <u>crack</u> under pressure.
The patient <u>felt</u> the tube <u>enter</u> his stomach.

Perception verbs differ from causative verbs in that their complement can also be in the continuous (V_{ing}) form. The continuous form emphasizes the process; the simple form emphasizes the result. Examples:

The engineer <u>saw</u> the oil <u>floating</u> near the platform.
The space scientists <u>are watching</u> the probe <u>passing</u> through the asteroid belt.
The terrified meteorologists <u>heard</u> the tornado <u>crashing</u> through the roof.
The doctor <u>felt</u> the tumor <u>pressing</u> against the patient's spine.
The astronomers <u>will observe</u> the sun <u>ejecting</u> a solar flare into space during the next eclipse.

EXERCISE 4.5 Combine the following sentences, using infinitive structures **without** *to*.

Example: We saw X. X=The moon rose.
 Answer: We saw the moon rise.

1. A fume hood lets X. X=Chemists combine noxious substances without inhaling them.
2. Geophysicists watched X. X=Lava streamed from the rift zone for twenty-four hours.
3. A rat growth hormone implanted into mice made X. X=The mice grew to twice their normal size.
4. Observers heard X. X=The thunder cracked exactly 3.2 seconds after the lightning flash.
5. Termites may X. X=Termites contribute as much as 50 percent of atmospheric methane, according to some researchers.
6. An earthquake of magnitude 8.3 on the Richter scale would X. X= An earthquake of magnitude 8.3 on the Richter scale will cause considerable damage in the San Francisco metropolitan area.
7. The government had X. X=The chemical company cleaned up the hazardous-waste site.
8. Patients do not feel X. X=A brain surgeon probes their exposed cerebral hemispheres.
9. The population should not X. X=The population eats fish from a dioxin-contaminated river.
10. The meteorologists carefully observed X. X=Tornado-like vortexes developed in the tornado simulator.

EXERCISE 4.6 Write the correct form of the verbs in parentheses in the blanks.

1. Space scientists would like (know) _____ if there are methane oceans on Saturn's moon Titan.
2. A seismograph allows a geologist (measure) _____ the vibrations within the earth.
3. A plant must (have) _____ good drainage (survive) _____ .
4. The doctor used a stethoscope (hear) _____ the patient's heart (beat) _____ .
5. Welded joints will (replace) _____ the rivets in the weakest section of the bridge.
6. In this electron micrograph, you can see the zygote (undergo) _____ mitosis.
7. A high-pressure area in the Northern Hemisphere makes air masses (spin) _____ in a clockwise direction.
8. After drinking a specific quantity of alcohol, the subject was asked (recall) _____ a list of words.

9. The nuclear-plant engineers did not notice the warning light (come on) ———————————— .

10. The gynecologist warned the patient that she had better not (have) ———————————— another child.

Section 5
WRITING AIDS

VERB-PHRASE PARALLELISM

In Unit II, we studied noun-phrase parallelism because it was particularly required in constructing the plan-of-development sentence in a description of a mechanism. Now we will look at verb-phrase parallelism because we will need it in constructing the plan-of-development sentence in a description of a process.

Types of Verb Phrases

A verb phrase consists of a verb and its modification. There are four types of verb phrases in English: (1) a finite verb (subject plus verb), (2) an infinitive verb with *to*, (3) an infinitive verb without *to*, and (4) gerund (V_{ing}) as a present participle.

TYPE	EXAMPLE
1. a finite verb	
a. nonmodified	The plant grows/grew/has grown new leaves. (active)
	The plant is grown in this area. (passive)
b. modified	The plant grows rapidly.
2. an infinitive + *to*	
a. nonmodified	The plant needs sun to grow. (active)
	The plant needs to be grown in the shade. (passive)
b. modified	The plant needs sun to grow rapidly.
3. an infinitive without *to*	
a. nonmodified	
1) with a modal	You should grow tomatoes in bright sunlight. (active)
	This plant should be grown in the shade. (passive)
2) with a causative verb	We let the plant grow for two weeks.
3) with a perception verb	We saw the plant grow new leaves.
b. modified	The plant will grow rapidly.
4. a gerund (V_{ing}) as a present participle	
a. modified with a prepositional phrase	Growing in sandy soil, the plant has no flowers. (active)
	Being grown in desert regions, the plant needs constant irrigation. (passive)
b. modified with an adverb	Growing rapidly, the plant is a good source of food for cattle.

EXERCISE 5.1   Underline the verb phrases in the following passage.

CIVIL ENGINEERING[6]

Civil engineering deals with the design and construction of objects that are intended to be stationary. It offers a particular challenge because every structure of a system that is designed and built by civil engineers is unique. One structure rarely duplicates another exactly. Even when structures seem to be identical, site requirements or other factors generally result in modifications. Large structures like dams, bridges, or tunnels may differ substantially from previous structures. The civil engineer must therefore always be ready and willing to meet new challenges. Within the field of civil engineering itself, there are subdivisions: structural engineering, which deals with permanent structures; hydraulic engineering, which is concerned with systems involving the flow and control of water or other fluids; and sanitary or environmental engineering, which concerns the study of water supply, purification, and sewer sytems.

Verb-phrase parallelism is required if two or more verb phrases occur in a single sentence. These verb phrases are commonly connected by the same words we used in noun-phrase parallelism (Unit II, Section 5), including (1) *and,* (2) *both . . . and,* (3) *whether . . . or,* (4) *either . . . or,* (5) *neither . . . nor,* (6) *not,* and (7) *not only . . . but (also).*

1. Gisela wants to study engineering <u>and</u> to practice medicine.
2. She feels she should <u>both</u> study engineering <u>and</u> practice medicine.
3. At first, she could not decide <u>whether</u> to study engineering <u>or</u> to practice medicine.
4. Her parents thought that she should <u>either</u> study engineering <u>or</u> practice medicine.
5. For a while, she thought that she would <u>neither</u> study engineering <u>nor</u> practice medicine.
6. Later, she thought that she should study engineering, <u>not</u> practice medicine.
7. Now, she <u>not only</u> studies engineering <u>but also</u> practices medicine.

EXERCISE 5.2   Complete the sentences with the correct parallel form of the sentence in parentheses.

Example: The plane will arrive at eight _____ . (The plane will leave at noon.)
  Answer: The plane will arrive at eight and leave at noon.

1. Materials engineers test the strength of materials and _____ (The material's resistance to corrosion is evaluated.)

   _____.

2. Biologists studying vent microbes will both increase ambient pressure _____

[6]Adapted from "Fleming," in *Compton's Pictured Encyclopedia* © 1964 by Encyclopaedia Britannica, Inc.

in an effort to determine the upper limit of conditions under which the bacteria can live. (The temperature will be raised to 500° C.)

3. The volcano was only _____, not ejecting lava. (Steam was being given off.)
4. Sanitary installations want neither to incinerate nuclear waste, because of toxic fumes, _____.
_____.
(Dumping nuclear waste is not acceptable, because of groundwater contamination.)
5. Environmentalists would prefer that nuclear waste be _____
_____.
or not created at all. (Decontaminate waste by means of a chemical process.)
6. Mission control did not know whether to try to send the satellite into a higher orbit _____.
(Should they allow the satellite to fall to earth?)
7. When a sugar maple is attacked by parasites, it can not only make itself unpalatable to the invader _____
_____.
(It warns neighboring trees to protect themselves.)
8. Porpoises navigate by using their eyes _____
_____.
(Sound is focused onto a special "sound lens" in their head.)
9. Digital displays require a constant source of electricity _____
_____ but also to keep the digits from disappearing. (The numbers must be changed.)
10. Using a radiograph tracer, the doctors watched the macrophage find the virus _____ (The virus was surrounded by the macrophage.)

REFERRING TO SEQUENTIAL DIAGRAMS

In describing a cycle or a process, we must often refer to a diagram or a pictorial representation of a sequence. In fact, if any diagram, graph, table, or other pictorial material is used, it **must** be referred to in the text. This is primarily because the reader cannot be expected to shift his or her attention to a diagram without instructions from the text.

Look at Figure 3.1 on page 120. The entire diagram can be referred to in different ways:

1. Figure 3.1 shows the production of oxygen from mercurous oxide.
2. Oxygen gas is produced by heating mercurous oxide
   a. as shown in Figure 3.1.
   b. as in Figure 3.1.
   c. (see Figure 3.1).
   d. (Figure 3.1).

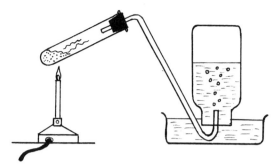

Figure 3.1 Production of O_2 from HgO

It is also possible to describe the steps in the process. The first step of the process shown in Figure 3.1 can be referred to in several different ways:

1. The HgO is heated in step A (or step 1).
2. In step A (or step 1), the HgO is heated.
3. The HgO is heated at A.
4. The HgO is heated (step A).
5. The HgO is heated (A).

The second step can be referred to in the same three ways:

1. The oxygen is collected in step B (or step 2).
2. In step B (or step 2), the oxygen is collected.
3. The oxygen is collected at B.
4. The oxygen is collected (step B).
5. The oxygen is collected (B).

EXERCISE 5.3   Fill the blanks in the following sentences with information from the three diagrams below.

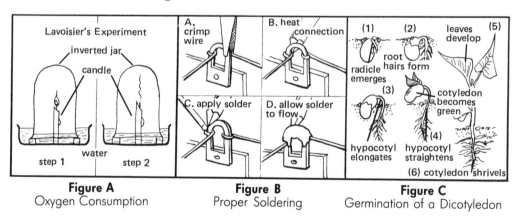

Figure A
Oxygen Consumption

Figure B
Proper Soldering

Figure C
Germination of a Dicotyledon

Example: A wire is crimped to the soldering lug ——————————— .
Answer: A wire is crimped to the soldering lug in Step A.

1. Lavoisier demonstrated that a burning candle consumes oxygen ——————————— .
2. The candle is lit ——————————— and covered with an inverted jar ——————————— .
3. The water rises inside the inverted jar ——————————— .
4. ——————————— the burning candle is extinguished.
5. The proper soldering of an electrical connection requires four steps, ——————————— .
6. ——————————— the wires are crimped to the soldering lug.
7. The connection is heated ——————————— .
8. Solder is applied ——————————— and allowed to flow ——————————— .
9. ——————————— shows the germination of a dicotyledon.
10. Root hairs emerge from the radicle ——————————— .
11. ——————————— , the cotyledons become green and the hypocotyl straightens.
12. Leaves develop ——————————— as the cotyledon shrivels.

REFERRING TO EQUATIONS

It is often necessary to use mathematical or chemical equations in scientific papers, particularly in production sequences. The most common way to use equations in text is as follows:

The general equation for the venturi meter is

$$\frac{V_A^2}{2g} + \frac{P_A}{\gamma} + H_A = \frac{V_B^2}{2g} + \frac{P_B}{\gamma} + H_B - Z_0 \tag{1}$$

where

V_A = velocity in Section A
g = gravity
P_A = pressure at Point A
γ = specific weight of the liquid
H_A = height in Section A
Z_O = loss of charge

Notice that the equation is centered alone on the page. The number to the right is necessary only if there is a sequence of equations. In that case, it is simpler later in the paper to refer to "equation 1." The elements of the equation are described directly below the equation, the first element preceded by the word *where*.

If the equation is very simple, it is sometimes included as part of a sentence, not centered on the page.

Einstein's equation $E = mc^2$, where E = energy in ergs, m = mass in grams, and c = the speed of light in cm/sec, ushered in the atomic age.

DESCRIBING SIMPLE MATHEMATICAL FUNCTIONS

The various forms of the four simple mathematical functions in the chart below are often required in descriptions of processes.

FUNCTION	SIGN	PROCESS	VERB	RESULT	TWO-WORD VERB
addition	+	plus	to add	the sum	to add X to Y
subtraction	−	minus	to subtract	the difference	to subtract X from Y
multiplication	×	times	to multiply	the product	to multiply X by Y
division	÷	divided by	to divide	the quotient (+ the remainder)	to divide X by Y

Look at the different ways that the following examples can be stated:

Example 1: $2 + 7 = 9$

a. Two plus seven $\begin{cases} \text{equals} \\ \text{is equal to} \\ \text{makes} \\ \text{is} \end{cases}$ nine.

b. If we add two to seven, $\begin{cases} \text{the answer is} \\ \text{we get} \end{cases}$ nine.

c. The sum of two and seven is nine.

d. $\begin{cases} \text{The addition of} \\ \text{Adding} \end{cases}$ two to seven $\begin{cases} \text{equals} \\ \text{is equal to} \\ \text{produces} \end{cases}$ nine.

Example 2: $6 \times 3 = 18$

a. Six times three $\begin{cases} \text{equals} \\ \text{is equal to} \\ \text{makes} \\ \text{is} \end{cases}$ eighteen.

b. If we multiply six times three, $\left\{ \begin{array}{c} \text{the answer is} \\ \text{we get} \end{array} \right\}$ eighteen.

c. The product of six and three is eighteen.

d. $\left\{ \begin{array}{c} \text{The multiplication of} \\ \text{Multiplying} \end{array} \right\}$ six by three $\left\{ \begin{array}{c} \text{equals} \\ \text{is equal to} \\ \text{produces} \\ \text{gives} \end{array} \right\}$ eighteen.

EXERCISE 5.4   Solve the following problems and write out the complete problem and answer in words. Choose a different method in each case so that you can practice all the forms.

Example: $24/6 = 4$
  Answer: Twenty-four divided by six equals four.

1. $11 + 2$
2. $116 - 39$
3. 44×44
4. $734/28$
5. $\frac{3}{4} + \frac{2}{3}$
6. $(20.019 \times 3.1416) - 7$
7. $4000/(2 \times 10^3)$ ($10^3 = $ "ten cubed")
8. $a^2 + b^2 = 18$; solve for a ($\sqrt{} = $ "the square root of")
9. $(49/7) + (22 \times 3)$
10. $\pi r^2 = 51$; solve for r ($\pi r^2 = $ "pi r squared")

PART II
Writing a Description
of a Process

Section 6
PREWRITING ACTIVITY: DESCRIBING CYCLES AND STEPS

A description of a process is used in scientific writing to explain how something works, how something is done, or how something was invented or discovered. Before we look more closely at describing a process, do the following exercise.

EXERCISE 6.1   Choose one of the "Cycles for Description" illustrated below. Several vocabulary words are indicated in the drawings, but there are no complete sentences. Write a paragraph describing the cycle you have chosen.

In order to understand a complete process, we must identify all its steps. The following exercise will help you to do this.

The Water Cycle

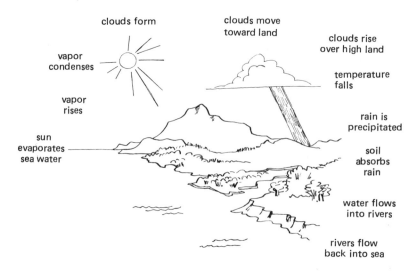

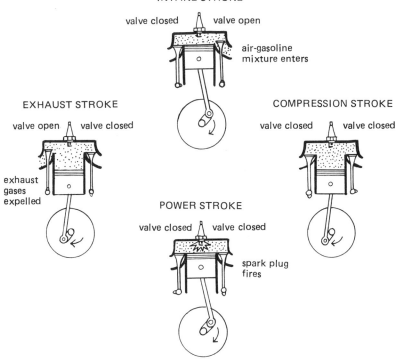

INTAKE STROKE

valve closed   valve open

air-gasoline
mixture enters

EXHAUST STROKE

valve open   valve closed

exhaust
gases
expelled

COMPRESSION STROKE

valve closed   valve closed

POWER STROKE

valve closed   valve closed

spark plug
fires

The Four-Stroke Engine

EXERCISE 6.2   1. Choose one of the processes from the list on page 126. Consider the process carefully and write down any aspect of the process that you can think of. Describe your results to your partner.

2. With your partner, decide which steps are not so important because they are obvious or belong to a step that you have already mentioned. Make sure that each major step is of equal weight (has the same degree of significance as the others). Arrange the steps in chronological order. Now do the same for your partner's list.

Example: how a photographic film is developed

1. Aspects of the process
   expose film   rinse film in stop bath
   mix chemicals   use a red light bulb
   immerse film in developer   immerse film in fixer
   find a darkroom   discard developer
   remove film from camera   dry film

2. Determining significant steps

Significant steps	Insignificant steps
a. immerse film in developer	expose film
b. rinse film in stop bath	remove film from camera
c. immerse film in fixer	mix chemicals
d. dry film	find a darkroom
	use a red light bulb
	discard developer

List:

How Something Works	How Something Is Done	How Something Was Discovered or Invented
1. how the immune system functions	7. how a foundation is constructed	13. how penicillin was discovered
2. how alcohol is produced	8. how age is determined	14. how the atomic bomb was invented
3. how cement solidifies	9. how oil is refined	15. why the big-bang theory was developed
4. how a telephone works	10. how a heart is transplanted	16. how electricity was discovered
5. how a star is formed	11. how computer data are stored	17. how the structure of DNA was determined
6. how flowering plants reproduce	12. how a frog is dissected	18. how the computer was developed

Section 7
STRUCTURE

A typical description of a process introduces the subject, describes the steps of the process in detail, and makes a conclusion. If the process has three steps, the description usually consists of five paragraphs: one for the introduction, one for each step, and one for the conclusion. Of course, if the paragraphs are very long, more than one might be required for one step. This is because one of the functions of a paragraph is to provide eye relief to the reader. Let us look at the structure of a process description in detail.

PARAGRAPH 1: INTRODUCTION

The introduction to a description of a process usually consists of three to four sentences in the following order:

1. Formal definition: What is the process?
2. Purpose: Why is/was this process performed? Why does/did it occur?
3. Scope (optional): From what point of view will this process be described?
4. Plan-of-development sentence: What are the chief steps or stages of the process?

How something is done can be explained in two ways: (1) as a description or (2) as instructions. Instructions differ from descriptions in that the imperative (command) form of verbs is used and the steps are usually listed. The writer must be careful not to change from a description to an instruction within a paragraph.

Formal Definition

A formal definition is the best way to begin a description of a process (see Unit I, Section 2). If appropriate, a second sentence can be added to indicate who performs this process.

Purpose of the Process

If the purpose of the process is not part of the formal definition, it can be expressed in sentences such as the following:

Example: developing a film

{ The reason for / The purpose of } developing a film is to produce a negative from which a photograph can be printed.

Films { are / must be } developed before a photograph { is / can be } made.

Example: the digestive system

{ The function / The purpose } of the digestive system is to make food available to the cells of the body.

Food { is / must be } digested before it { is / can be } used by the body.

Scope (optional)

The scope or point of view of the description indicates the appropriate audience. Using film developing as an example, we can imagine a variety of possible readers:

1. a person looking for general instructions about photography
2. a camera-store owner trying to determine which chemicals should be ordered
3. an engineer trying to develop a machine that will automate the process of film developing
4. a customer trying to prove that a damaged film was the developer's fault
5. a biologist interested in the commercial application of acetic acid

In general, the scope sentence indicates the degree of detail to be expected. This can also be indicated by the title of the description (e.g., "Photography for the Beginner," "The Chemistry of Film Developing") or by the nature of the language in the introduction (highly technical or nontechnical). Here are some example scope sentences:

The process to be described will focus on $\Big\}$ . . .
The purpose of this description is to explain

. . . $\left\{\begin{array}{l}\text{the financial aspects of}\\\text{the chemicals required for}\\\text{the basic technique of}\end{array}\right\}$ film developing.

Plan-of-Development Sentence

The plan-of-development sentence is the most important sentence in the entire description. It tells the reader the order of presentation of the different steps or stages in the process. There are several ways to write this sentence:

X consists of A, B, and C.
X consists in A, B, and C.
X requires A, B, and C.
X involves A, B, and C.
X involves the following steps: A, B, and C.
X has the following steps: A, B, and C.

Example: developing a film

Film developing consists of
Film developing consists in
Film developing requires $\left.\begin{array}{l}\\\\\\\\\\\\\end{array}\right\}$ $\left\{\begin{array}{l}\text{developing, stopping,}\\\text{fixing, and drying the film.}\end{array}\right.$
Film developing involves
Film developing involves the following steps:
Film developing has the following steps:

EXERCISE 7.1   Look at the steps (or stages) you determined in Exercise 6.2. Form them into a plan-of-development sentence using one of the model sentences above.

PARAGRAPH 2: DESCRIPTION OF STEP A

The second paragraph in a description of a process describes the first step or stage. The first sentence identifies the step: What is it? The second sentence describes the purpose of the step, if this has not already been stated: Why is/was it performed? Why does/did it occur? The third and subsequent sentences describe the substeps in detail using chronological (time) order (see Unit

IV, Section 2, "The TIME Group"). It is often necessary to describe a part of a mechanism in describing a process. Some writers prefer first to describe the part in detail and then to describe the substeps that concern that part. Other writers describe a subpart followed by a substep of the process, then another subpart followed by the next substep of the process. (We also saw in Unit II that in a description of a mechanism, the process can be explained in the conclusion.)

PARAGRAPH 3: DESCRIPTION OF STEP B

Paragraph 3 is constructed in the same manner as paragraph 2.

PARAGRAPH 4: DESCRIPTION OF STEP C

Paragraph 4 is constructed in the same manner as paragraph 3.

PARAGRAPH 5: CONCLUSION

Most formal compositions have a concluding paragraph. In a description of a process, there are several ways to write a conclusion:

1. Describe the advantages of the process.
2. Describe the disadvantages of the process.
3. Describe the specialized uses and applications of the process.
4. Describe the most recent development in the process, the latest refinement of the process, or the most recent discoveries about the process.

If there is not enough information to explain each step of the process in detail, the entire process may be described chronologically in the second paragraph. Some writers prefer to list the steps of the process by number to avoid using sentence connectors of time (e.g., *then, next, after that*), especially in instructions. If this is done, the plan-of-development sentence is often deleted and the numbers are indented like this:

> XXXXXXXXXXXXXXXXXXXXXXXXXXXXXXXXXXXXXXXXXXXX
> XXXXXXXXXXXXXXXXXXXXXXXXXXXXXXXXXXXXXXXXXXXX
> XXXXXXXXXXXXXXXXXXXXXXXXXXXXXXXXXXXXXXXXXXXX
> XXXXXXXXXXXXXXXXXXXXXXXXXXXX.
> 1. XXXXXXXXXXXXXXXXXXXXXXXXXXXXXXXXXXXXXXXXXX
>   XXXXXXXXXXXXXXXXXXXXXXXXXXXXXXXXXXXXXXXXXX
>   XXXXXXXXXXXXXX.
> 2. XXXXXXXXXXXXXXXXXXXXXXXXXXXXXXXXXXXXXXXXXX
>   XXXXXXXXXXXXXXXXXXXXXXXXXXXXXXXXXXXXXXXXXX
>   XXXXXXXXXXXXXXXXXXXX.
> 3. (etc).

Section 8
MODELS

MODEL 1: TAKING A BLOOD SAMPLE

Taking a blood sample is a procedure by means of which blood is removed from a patient for analysis. It should be done by a nurse or other trained medical practitioner so that infection may be avoided. Before taking the sample, the nurse should have a tourniquet, some cotton balls soaked with alcohol, a needle, and a syringe. There are two steps in taking the sample: preparing the patient and drawing the blood.

The tourniquet is placed on the patient's arm about three inches above the elbow. The patient is requested to make a fist with the arm extended. The vein, which protrudes because of the tourniquet, can be located with a finger. The area around the vein should be wiped with a cotton ball soaked with alcohol.

The needle, which has been inserted into the syringe, is carefully injected into the vein. Care should be taken not to pass completely through the vein. The plunger of the syringe is pulled back until the required amount of blood has been taken. The tourniquet is then loosened. A cotton ball soaked with alcohol is placed over the needle and the vein as the needle is pulled out. The patient is told to bend the elbow in order to prevent unnecessary bleeding. The blood is placed in a test tube labeled with the patient's name.

This method of taking blood is relatively painless and is recommended when a substantial quantity (more than a drop) of blood is required.

MODEL 2: THE WANKEL ENGINE[7]

The Wankel engine is a form of heat engine which has a rotary piston. In other words, instead of going up and down, the Wankel piston rotates in the cylinder. Both cylinder and piston are quite different in shape from those of conventional engines.

The Wankel piston is triangular with curved sides and the cylinder is roughly oval in shape. The piston has an inner bore which is linked through an eccentric gear to the output shaft. The other end of the bore is toothed and engages with a stationary gear fixed to the cylinder end. This arrangement ensures that the piston follows an elliptical path around the cylinder so that the apexes of the piston, which carry gas-tight seals, are always in contact with the inside surface of the cylinder.

The piston thus forms three crescent-shaped spaces between itself and the cylinder wall, which vary in size as the piston rotates. Fuel enters the cylinder through the inlet port when one of these spaces is increasing in size. The fuel trapped in this section is then compressed by the turning piston and ig-

[7]Eric H. Glendinning, *English in Mechanical Engineering* (Oxford: Oxford University Press, 1974), pp. 101–2.

nited by the spark plug. The expanding gases subject the piston to a twisting moment which makes the piston revolve further until the exhaust gases escape through the exhaust port. A fresh charge is then induced into the cylinder. Meanwhile, the same process is being repeated in the other two spaces between the piston and the cylinder.

The Wankel engine has many advantages over the reciprocating piston engine. Fewer moving parts are necessary because it produces a rotary movement without using a connecting rod and a crankshaft. Because of this rotary movement it has no vibration. In addition, it has no valves, it is smaller and lighter than conventional engines of the same power, and it runs economically on diesel and several other fuels.

Section 9
ANALYSIS

The basic structure of a description of a process, which we discussed in Section 7, is as follows:

     I. Introduction
        A. Formal definition
        B. Purpose
        C. Scope (optional or implied)
        D. Plan-of-development sentence
    II. Description of step A
   III. Descriptionof step B
   IV. Description of step C
    V. Conclusion

This structure can be varied considerably. If the description is a set of instructions, the plan-of-development sentence is often deleted, especially if the steps are numbered so that the plan of development is immediately clear. Instructions often have no conclusion, since the conclusion will be the result of the instructions.

If the series of steps in the description cannot be readily broken down into substeps, the entire description of the process may be written in one paragraph. Another paragraph may describe the mechanism required for the process. Thus, another structure for the description of a process may be as follows:

     I. Introduction
        A. Formal definition
        B. Purpose
    II. Description of the mechanism
   III. Description of the process
   IV. Conclusion

EXERCISE 9.1   With another student, preferably one who is in your field, analyze Model 1 in Section 8. Determine the function of each sentence in the model (formal definition, plan-of-development sentence, chronological details, conclusion, etc.), and label each sentence in the margin. Discuss your results with your partner. Do you agree?

EXERCISE 9.2   Analyze Model 2 in the same way as in Exercise 9.1. Compare your analysis with that of Model 1. Are they different? If so, how?

Section 10
CHOOSING A TOPIC

Your assignment is to write a description of a process in your own field. First you must choose a topic. Many examples have been given in this unit that are good topics for a description of a process (e.g., the discovery of a new medicine, the formation of a star, the construction of a foundation). Here are some additional topics related to certain majors:

Biology/Plant Physiology	Civil/Mechanical Engineering
Contouring Land	Constructing a Dam
Glycolysis	The Disposal of Sludge
Measuring Fitness in Evolution	A Gyroscopic Stable Platform
The Swedish Increment Borer	Reinforcing Concrete

Electrical Engineering/Computer Science	Chemistry/Chemical Engineering
The Compilation Process	Combustion
The Fabrication of Microelectronic Circuits	Determining a Melting Point
	Gel Chromatography
Installing a Power Transformer	The Production of Gasohol
The Transmission of Television	

Medicine/Physiology	Physics/Astronomy
The Artificial Lung	Accelerating Electrons
The Hearing Mechanism	The Formation of the Universe
The Regeneration of Epithelial Membrane	Gravity
	How Ocean Waves are Formed
The Transmission of Eucaryotic Chromosomes	

EXERCISE 10.1   Write a description of a process in your own field, using one of the structures shown in Section 9.

UNIT IV
The Classification

PART I
Grammar

Section 1
ARTICLES: THE GENERIC ARTICLE

GENERAL DESCRIPTION OF GENERIC NOUN PHRASES

There are two major classifications of noun phrases in English: (1) specific noun phrases and (2) generic noun phrases. Specific noun phrases refer to actual objects, people, quantities, or ideas. In Units I, II, and III, we focused on the articles associated with specific noun phrases, including the concept of first and second mention and shared knowledge.

Generic noun phrases refer to symbolic or representative objects, people, quantities, and ideas. Their purpose is not to show a specific example but rather to show what is normal or typical for the members of a class. For this reason, definitions always use the generic article, as we shall discuss in a moment.

Let us compare some specific noun phrases with some generic ones:

CLASSIFICATION	SPECIFIC (first mention)	GENERIC
Singular Countable	John bought <u>a calculator</u> yesterday.	<u>A calculator</u> is a useful instrument.
Plural Countable	John bought (some) <u>books</u> yesterday.	<u>Books</u> are valuable things.
Uncountable	Add (some) <u>water</u> if necessary.	<u>Water</u> is composed of hydrogen and oxygen.
	SPECIFIC (second mention)	GENERIC
Singular countable	<u>The calculator</u> was not expensive.	<u>The atom bomb</u> has changed our lives.
Plural countable	<u>The books</u> John bought are chemistry texts.	- - - - - - - - - - - - - - - - - - - -
Uncountable	<u>The water in this river</u> is polluted.	- - - - - - - - - - - - - - - - - - - -

Notice that the specific sentences are concerned with **actual** things, whereas the generic sentences are concerned with **symbolic** or **representative** things.

EXERCISE 1.1   Indicate whether the underlined phrase in the following sentences is generic (G) or specific (S).

Examples: ___ <u>The sun</u> is a medium-sized star.
Answer: _S_ <u>The sun</u> is a medium-sized star.

1. ———— <u>The lion</u> lives in Africa.
2. ———— <u>The computer</u> is down today.
3. ———— <u>The neutron bomb</u> was carefully placed in the plane.
4. ———— The nurse replaced <u>the patient's bandage</u>.
5. ———— <u>The lion</u> is lying in its cage.
6. ———— Acceptance for publication depends on <u>the significance of the study</u>.
7. ———— The newest member of the nuclear arsenal is <u>the neutron bomb</u>.
8. ———— <u>The computer</u> is affecting the way we live.
9. ———— <u>A bandage</u> protects wounds while they are healing.
10. ———— <u>The significance of the study</u> was that it proved the existence of volcanoes in outer space.

Since generic noun phrases tell us general information about groups or classes of things, they are usually found with simple verb forms and not with the continuous tenses:

Incorrect:
*Tigers (as a class) are living in Asia.
*Tigers (as a class) were living in Asia.
*Tigers (as a class) will be living in Asia.

Correct:
Tigers (as a class) live in Asia.
Tigers (as a class) lived in Asia.
Tigers (as a class) have lived in Asia.
Tigers (as a class) had lived in Asia.
Tigers (as a class) will live in Asia.

However, a continuous form is possible with a generic noun phrase if it implies cause and effect:

The computer <u>is becoming</u> a fact of life. (effect: "computer society")
The computer <u>is taking</u> the place of workers. (effect: unemployment)
The computer <u>is helping</u> some students to improve their mathematical skills. (effect: better learning)

EXERCISE 1.2   Decide whether the subject of the verb in parentheses is generic or specific and draw a line through the incorrect verb. In some cases, both forms are correct.

Example: Solder (contains/is containing) mostly lead and tin.
 Answer: Solder (contains/~~is containing~~) mostly lead and tin.

1. Water (is consisting/consists) of hydrogen and oxygen.
2. The dentist (is sharpening/sharpens) her instruments for tomorrow's patients.
3. "Yellow rain" is a toxin that (is being found/is found) in cereal grains.
4. The dinosaur, as a class, (was living/lived) at the time when flowering plants appeared.
5. Some researchers have found that vitamin C (blocks/is blocking) the beneficial effect of copper in the body.
6. The depth of a burn (is being measured/is measured) by means of ultrasound.
7. CO_2 in the atmosphere (is affecting/affects) the temperature of the earth.
8. In this picture, the cell (is undergoing/undergoes) mitosis.
9. These seeds (are not germinating/do not germinate) because the soil is too wet.
10. Nowadays optical circuitry (is competing/competes) with electronic circuitry in computers.

The generic articles can be divided into two groups: abstract generic articles and concrete generic articles.

THE ABSTRACT GENERIC ARTICLE: *the*+A Singular Countable Noun

The abstract generic article *the* refers to the class itself and never to the representatives of that class. For this reason, it occurs only with a singular noun (the name of the class), which must be countable. Look at this example:

The class itself:	*Representatives of the class:*
The lion lives in Africa.	A lion lives in Africa.
	(∅) Lions live in Africa.
(The lion = the entire class of animals called lions.)	(A lion/Lions = examples of the class of animals called lions.)

Abstract generic noun phrases occur in two contexts: (1) in sentences that imply cause and effect and (2) in definitions.

Sentences That Imply Cause and Effect

We have already seen that nouns that imply cause and effect are the only type of generic nouns to occur with continuous tenses. Causes include diagnoses and purposes; effects include results, solutions, and inventions:

The herpes virus has affected 20 million Americans. (diagnosis)
The stethoscope improved the diagnosis of heart problems. (purpose)
The cyclotron opened the field of particle physics. (results)
The smallpox vaccine has practically eradicated the disease throughout the world. (solution)
The world seems smaller because of the telephone. (invention)

Definitions

Definitions include classifications, attributes, comparisons, and any other means we use to define a species. In Unit I, we learned Aristotle's definition formula: An A is a B that C. Now we see that it is possible to say, "The A is a B that C" **if** A is a class, not a representative:

The coyote (is an animal that) is a useful predator. (classification)
The neutron bomb (is a weapon that) destroys people but not property. (attribute)
The abacus is sometimes as fast as the computer. (comparison)

Notice that in a definition, it is usually only the A section that has abstract generic *the*, not the B section, except in comparisons:

*The automobile is powered by the internal-combustion engine.
The automobile is powered by an internal-combustion engine.

EXERCISE 1.3   Keeping the same word order, use the following information to make complete sentences using abstract generic *the*.

Example: thermometer/instrument/measure/temperature
 Answer: The thermometer is an instrument that measures temperature.

1. optic nerve/connect/eyes/to brain
2. Marconi/invent/radio
3. evergreen/be/tree/stay green all year round
4. venturi meter/measure/rate of flow
5. internal-combustion engine/lead to/inexpensive forms of transportation for the general population
6. heart-lung machine/allow/doctors/repair or replace/these organs
7. amoeba/cause/some forms of dysentery
8. fluorescent light bulb/be more efficient than/incandescent light bulb
9. multiple-arch dam/be suitable for/remote locations
10. Geiger counter/detect/ionizing radiation

Abstract generic nouns in discourse (i.e., text) are different from specific nouns in that, since they are **symbolic** rather than actual, they do not follow the rule of first and second mention. Look at the following examples:

Specific (first/second mention):

An old man is walking with a small boy. The man is tired,

but the boy is hungry and needs some food.

Abstract generic (no first/second mention):
The noblest bird of prey is the eagle. The eagle feeds on a variety of small animals. The eagle has been put on the endangered-species list. It would be sad to lose the eagle just because we failed to take adequate steps to protect it.

However, the first/second-mention concept does occur in concrete generic contexts, especially in descriptions of processes:

Nuclear reactors generate heat. The heat is transferred to water. The water becomes steam. The steam drives a turbine.

The only way to distinguish an abstract generic from a specific second-mention noun phrase is by context:

Abstract generic:
The neutron bomb is a devastating weapon.

Specific second mention:
Please remove the neutron bomb from my car.

EXERCISE 1.4   Add generic *the* to the following passage where necessary.

Analysis: Can you find any examples of first/second mention?

THE RATTLESNAKE[1]
Rattlesnake is one of the four poisonous snakes in North America. It belongs to family of pit vipers, *Crotilidae*. There is a deep pit on each side of face, between eye and nostril. The pit contains a heat-sensitive membrane that helps rattlesnake find its warm-blooded prey. Although rattlesnake is a reptile, female bears live young; that is, the eggs hatch inside mother's body. Rattle from which snake's name is derived consists of a series of rings that strike together when snake is excited.

THE CONCRETE GENERIC ARTICLE

Concrete generic noun phrases refer to the representatives of a class and never to the class itself. The concrete generic article occurs with (1) singular countable nouns and (2) plural countable nouns and uncountable nouns.

Singular Concrete Generic: *a(n)*

The singular concrete generic article is *a(n)*. Like the abstract generic, which can only be singular, the singular concrete generic also occurs in two contexts: (1) when a generalized instance rather than the actual noun is meant, and (2) in definitions.

Generalized instance.   When a generalized instance rather than the actual noun is meant, the term *generalized instance* signifies the idea, archetype, or form of an object. Look at this example:

Concrete generic:
A physics student needs a calculator.

The underlined words both indicate generalized instances: any person who studies physics has particular needs, which include devices for rapid calculation. Notice how this sentence is different from specific first mention:

Specific first mention:
A physics student was trying out a calculator in the bookstore when the robbery took place.

The underlined words indicate an actual student, not just the idea of a student, and an actual calculator, not just the idea of a calculator.

[1]Adapted from "Rattlesnake," in *Compton's Pictured Encyclopedia*, © 1964 by Encyclopaedia Britannica, Inc.

Here are some other examples of concrete generic noun phrases with generalized instances:

> An airplane can seed a cloud without silver iodide.
> A satellite photograph helps a scientist to understand the geological structure of the earth.

EXERCISE 1.5 Indicate whether the noun phrase following the blank is generic (G) or specific (S). Then add *a* or *an* to the blanks.

Example: _____ _____ flower attracts birds and insects.
Answer: _G_ _A_ flower attracts birds and insects.

1. _____ _____ ocean wave sometimes travels for hundreds of miles.
2. _____ _____ ocean wave destroyed the Sea Cliff Restaurant.
3. _____ _____ elephant eats an average of eighty pounds of feed a day.
4. _____ The lumber company has trained _____ elephant to move trees and logs.
5. _____ Surgeons found _____ tumor in the patient's lung yesterday.
6. _____ _____ tumor is generally removed as soon as possible.
7. _____ _____ carbon-carbon single bond is longer than _____ carbon-carbon double bond.
8. _____ The group determined that their labeled ethane molecule had _____ carbon-carbon single bond of 1.54 angstroms.
9. _____ The meter showed _____ voltage spike of 1200 volts.
10. _____ _____ voltage spike can damage solid-state components.

Definitions. Definitions belong to both the abstract **and** the concrete generic because we can define either the class (abstract generic) or the representatives of the class (concrete generic):

> The elephant is an animal that has a trunk. (class)
> An elephant is an animal that has a trunk. (representative)

Singular countable concrete generic nouns cannot be used with an implied cause and effect:

> Incorrect:
> *The world has become smaller because of a telephone.
> *A computer is changing our society.

These sentences are incorrect because they suggest that a **single** telephone or a **single** computer has brought about these effects. They require abstract generic *the* (see the preceding discussion of the abstract generic article).

EXERCISE 1.6   Correct the underlined parts of the following sentences if necessary. There are **no second-mention** examples.

Example: A telephone is connecting millions of homes and businesses all over the world.

Answer: The telephone connects millions of homes and businesses all over the world.

1. A termite is generating 15 to 30 million tons of atmospheric methane per year.
2. Every house should have the water supply in case of emergencies.
3. The light is traveling at 299,792,458 meters per second.
4. A snake was lying across the road.
5. An atomic bomb was invented in 1942.
6. Everyone should have the blood test before the age of twenty.
7. High-power accelerators are releasing short, high-power pulses.
8. Many people who live near rivers have installed the water wheel to provide electricity.
9. The oxygen in the atmosphere makes up 21 percent of the total.
10. A snake is defending itself with its teeth.

Plural Countable and Uncountable Concrete Generic Ø

The plural countable and uncountable generic article is Ø. Uncountable nouns and indefinite plural countable nouns already have a generic quality because they refer to a general mass or quantity and thus make the distinction between class and representative less clear. For this reason they are not limited to certain contexts, as are the singular forms we have discussed; they can occur in contexts of cause and effect, generalized instance, and definition.

The only difference between the specific and generic forms for plural countable and uncountable nouns is the implied or stated use of unstressed *some*. Unstressed *some* means "a portion or small quantity." (Stressed *some* means "a particular group or part that is different from 'the others," as in "Some elements are metallic; others are not.") Look at these examples:

Specific	Concrete Generic
Drink some milk.	Milk is good for you. (generalized instance)
Some steam was leaking from the pipe.	This engine runs on steam. (cause and effect)

They dropped some <u>nails</u>. <u>Nails</u> are metal spikes for holding things together. (definition)

The rule of unstressed *some* can help you to determine if a plural countable or uncountable noun is specific or generic.

EXERCISE 1.7 Indicate whether the underlined noun phrase in the following sentences is (G) generic or (S) specific.

Example: _____ The class performed <u>experiments</u> in the lab.
 Answer: _S_ The class performed <u>experiments</u> in the lab.

1. _____ The fire started because of <u>gasoline</u> on the floor.
2. _____ <u>Earthquakes</u> are violent movements in the earth's crust.
3. _____ My brother is studying <u>medicine</u> at night school.
4. _____ <u>Gasoline</u> is a highly flammable, volatile hydrocarbon.
5. _____ This lab needs new <u>glassware</u>.
6. _____ The geologists felt <u>earthquakes</u> before and after the eruption.
7. _____ New <u>glassware</u> is quite expensive.
8. _____ You must take <u>medicine</u> if you want to get better by tomorrow.
9. _____ <u>Bird feathers</u> clogged the air intake of the jet engine.
10. _____ <u>Bird feathers</u> are typically light, hollow, and water-resistant.

The only way to change a concrete generic noun phrase into an abstract generic one is to make the noun phrase singular and countable, because it must refer to the class alone and not to the representatives of that class:

Concrete generic:
<u>Water</u> is composed of hydrogen and oxygen.
<u>Boll weevils</u> destroy cotton.
Abstract generic:
Incorrect:
*<u>The water</u> is composed of hydrogen and oxygen.
*<u>The boll weevils</u> destroy cotton.
Correct:
<u>The water molecule</u> is composed of hydrogen and oxygen.
<u>The boll weevil</u> destroys cotton.

EXERCISE 1.8 Label the underlined words in the following sentences abstract generic (AG) or concrete generic (CG). At the end of the sentence, indicate whether the context of the sentence is cause-and-effect (CAUS), definition (DEF), or generalized instance rather than actuality (GI).

Example: _____ <u>A whale</u> is a large marine mammal.   _____

Answer: __CG__ <u>A whale</u> is a large marine mammal.   __DEF__

AG/CG   Context

1. _____ <u>The cow</u> is a domestic animal.   _____
2. _____ Every farmer should have <u>a cow</u>.   _____
3. _____ <u>Cows</u> give us milk.   _____
4. _____ <u>Blood</u> transports oxygen and hormones.   _____
5. _____ <u>Mosquitoes</u> suck blood.   _____
6. _____ Sickle-cell anemia affects <u>the red blood cell</u>.   _____
7. _____ <u>The DNA molecule</u> was the subject of
           the investigation.   _____
8. _____ <u>A DNA molecule</u> encodes genetic information._____
9. _____ <u>DNA molecules</u> are found only in chromosomes._____

EXERCISE 1.9   Underline the generic noun phrases in the following passage. Indicate whether each phrase is abstract generic (AG) or concrete generic (CG).

Example: The human body has more than 500 muscles.

                 AG   CG
Answer: <u>The human body</u> has more than 500 <u>muscles</u>.
Analysis: Are there any examples of the first/second-mention effect?

Muscles[2]

The muscle is responsible for all body movement. By contracting and relaxing, muscles move legs and arms, turn eyes, and make faces smile or frown. Heart muscles pump blood through the circulatory system. Tongue and jaw muscles help us to chew food. The digestive system contains strong muscles that churn and move food.

Muscles also release heat. The heat keeps the body at a normal temperature. When it is cold, the body produces a shiver. This generates quick heat.

The muscle is made up of threadlike cells or fibers. These are grouped in bundles and wrapped in thin fibrous tissue. Muscle fibers are linked to the nervous system by motor nerves. When nerve impulses reach the fibers, they respond by contracting, becoming shorter and thicker. When the nerve impulses stop, the fibers relax, returning to their thinner, longer shape.

[2]Adapted from "Muscles," in _Compton's Pictured Encyclopedia_ © 1964 by Encyclopaedia Britannica, Inc.

Section 2
SENTENCE COMBINING

COORDINATION AND SUBORDINATION

Before we begin our discussion of coordination and subordination we need to understand the difference between a clause and a phrase. A clause is a part of a sentence that has a subject and a verb. A phrase is a part of a sentence that has no verb (or no subject, in the case of a verb phrase). Look at these examples:

Clauses		Phrases	
Relative clause:	which the insects ate	Prepositional phrase:	in the laboratory
Embedded question:	what the doctor said	Noun phrase:	a medical dictionary
Noun clause:	that the sun evolves	Verb phrase:	must have quickly evaporated

General Description

Coordination and subordination are the two major ways in which sentences are combined in English. If we want to show that two clauses have equal importance, we join them with coordinators. These include (1) the conjunctions (e.g., *and, but, yet, or, for,* and *so*) and (2) the conjunctive adverbs, also known as sentence connectors (e.g., *in addition, however, therefore*). Look at these examples:

	The beaker fell.
+	It did not break.

(1) The beaker fell, <u>but</u> it did not break.
(2a) The beaker fell; <u>however</u>, it did not break.
(2b) The beaker fell. <u>However</u>, it did not break.

The sentence connectors (conjunctive adverbs), like many other adverbs in English, can take several positions.

1. Before subject: John studies math; <u>however</u>, Mary studies law.
2. After subject: John studies math; Mary, <u>however</u>, studies law.
3. End of sentence: John studies math; Mary studies law, <u>however</u>.

It is much more common in English to show that one clause is **more** important than another. To do this, we join the clauses with clause subordinators. The clause subordinators include (3) the relative pronouns (see Units I, II and III, Section 2) and (4) the subordinating adverbs (e.g., *although, because, when*). Look at these examples:

(3) The beaker <u>that</u> fell did not break.
(4a) <u>Although</u> the beaker fell, it did not break.
(4b) The beaker did not break <u>even though</u> it fell.

We can also show that a clause is more important than a noun phrase in the same sentence. To do this, we join the clause and the noun phrase with a phrase subordinator. The phrase subordinators include (1) the complex prepositions, (e.g., *in spite of, in case of, due to*) and (2) other words (e.g., *not to mention, barring, until*).

(5) The beaker did not break <u>despite</u> its fall.

The coordinators and the subordinators (except the relative pronouns, which will not be discussed here) all indicate a particular logical relationship between each of the combined structures. However, as you can see in examples 1–5 above, the position of coordinators and subordinators is different, as is the punctuation of the sentences. Coordinators join two **independent** sentences. Clause subordinators require a **dependent** clause in the adverbial position of an **independent** sentence. Phrase subordinators require a noun phrase in the adverbial position of an **independent** sentence:

SUBJECT	VERB	ADVERBIAL
The beaker	did not break	<u>even though it fell</u>.
The beaker	did not break	<u>despite its fall</u>.

In the following descriptions, the numbers after each sentence connector indicate the position it can take—for example, *HOWEVER* (1,2,3). Capitalized words indicate the most frequently used words and phrases. The logical symbols indicate the formal logical relationship that the particular group of words shows.

Categories

Coordinators and subordinators can be divided into four major groups, based on the logical relationships they show: (1) the AND group, (2) the BUT group, (3) the SO group, and (4) the TIME group.

THE *AND* GROUP

Several words show the AND relationship between two sentences. The different forms of AND include addition, explanation, condition, listing, choice, summation, and transition.

Addition: ☐ and ☐ (Logical symbol: A.B)

Addition words indicate (1) more information, (2) surprising or unexpected information, or (3) a general or specific statement about the previous sentence.

1. Coordinators:

Different idea	Same idea
IN ADDITION (1)	*SIMILARLY* (1,2)
also (1,2)	*likewise* (1,2)
moreover (1,2)	*in the same way* (1)
furthermore (1,2)	

Examples: Dams control floods; in addition, they provide farmers with irrigation.
The moon orbits the earth; similarly, the earth orbits the sun.

2. Clause subordinators:

Different idea	Same idea
in addition to the fact that	*in the same way as*
	much as

Examples: Dams control floods in addition to the fact that they provide farmers with irrigation.
The moon orbits the earth in the same way as the earth orbits the sun.

3. Phrase subordinators:

Different idea	Same idea
IN ADDITION TO	- - - - - - - -
as well as	

Example: Dams control floods in addition to providing farmers with irrigation.

1. Coordinators:

Formal	Informal
FURTHERMORE (1,2)	*BESIDES* (1)
moreover (1,2)	*what is more* (1)
in fact (1,2)	*not only that* (1)

Example: The rain ruined the crops; furthermore, it washed out several bridges.

2. Clause subordinators:

not to mention the fact that

Example: The rain ruined the crops, not to mention the fact that it washed out several bridges.

3. Phrase subordinators:

not to mention

Example: The rain washed out the highway, <u>not to mention</u> several bridges.

General or specific information about the previous sentence.

1. Coordinators:

General	Specific
IN FACT (1,2)	*IN FACT*
indeed (1,2)	*indeed*
as a matter of fact (1,2)	*as a matter of fact*
in general (1,2)	*to be specific* (1,2)
generally speaking (1,2)	*in particular* (1,2)

Examples: Bromine replaces a hydrogen atom: <u>in fact</u>, all halogenation reactions occur in this way.

Halogens are added to the compound; <u>to be specific</u>, a chlorine replaces a hydrogen atom.

Research is being done in this area; <u>as a matter of fact</u>, Dr. Kalil is starting a new project next week.

2. Clause subordinators:
- - - - - - - - - - - - - - - - - -

3. Phrase subordinators:
- - - - - - - - - - - - - - - - - -

Explanation: □=□ (Logical symbol: A=B)

Explanation words indicate that the idea in the first sentence is defined or made clearer in the second sentence.

1. Coordinators:

IN OTHER WORDS (1,2,3)
that is to say (1,2)
that is (1)
i.e. (1) [from Latin *id est,* meaning "that is"]
to be precise (1,2)
put more simply (1)

Example: The tree is deciduous; <u>in other words</u>, it loses its leaves every year.

2. Clause subordinators:

by which (we) mean that
by which is meant that
the meaning being (that)

Example: The tree is deciduous, <u>by which is meant that</u> it loses its leaves every year.

3. Phrase subordinators:

- - - - - - - - - - - - - - - - - -

Condition: If □ then □ [□, or □] (Logical symbol: A⊃B)

Condition words indicate that the second sentence will or will not occur if the first sentence is true.

1. Coordinators:

Positive	Negative
in that case (1,2)	*OTHERWISE* (1)
in this case (1,2)	*if not* (1)
in other words (1,2,3)	

Examples: The fault slips; <u>in that case</u>, an earthquake occurs.
The pressure is released; <u>otherwise</u>, the reactor will explode.

2. Clause subordinators:

Positive	Negative
IF[3]	*UNLESS*
in the event that	
provided (that)	
as long as	
whether or not (irrespective of the condition)	

[3]Conditional sentences with *if* must commonly take one of three forms: 1) the real, 2) the present unreal, and 3) the past unreal conditional.
1. The real conditional describes a non-hypothetical situation. It indicates either a) one-time or b) general time.
a. one time: $if + V_{simple\ present,}\ V_{will}$
   If it <u>snows</u>, the roads <u>will be closed</u>.
b. general time: $if + V_{simple\ present,}\ V_{simple\ present}$
   If it <u>snows</u>, the roads <u>are closed</u>.
2. The unreal present conditional describes a hypothetical situation in the present:
   $if + V_{simple\ past,}\ V_{would}$
   If it <u>snowed</u>, the roads <u>would be closed</u>.
3. The unreal past conditional describes a hypothetical situation in the past:
   $if + V_{past\ perfect,}\ V_{would\ have}$
   If it <u>had snowed</u>, the roads <u>would have been closed</u>.

Examples: <u>If</u> the fault slips, an earthquake occurs.

The reactor will explode <u>unless</u> the pressure is released.

    3. Phrase subordinators:

Positive	Negative
in case of	*barring*
in the event of	

Examples: The sprinklers are activated <u>in the event of</u> fire.

The old building will stand, <u>barring</u> a strong earthquake.

Listing: ☐ ☐ ☐ ☐

    Listing words indicate a series of words or sentences that support a point or give examples.

    1. Coordinators:

FIRST (1) *SECOND* (1) *THIRD* (1)
one (1) *two* (1) *three* (1)
in the first place (1) *in the second place* (1)
furthermore (1,2) *moreover* (1,2)
(even) more importantly (1,2) *finally* (1,2)
for example (1,2) *for instance* (1,2)

Example: The space mission should not continue. <u>In the first place</u>, the rocket has not been adequately tested. <u>Moreover</u>, the outer surface of the capsule might burn up in the atmosphere. <u>Even more importantly</u>, the astronauts' air supply malfunctioned twice on the last mission.

    2. Clause subordinators:

FOR EXAMPLE
for instance
e.g. [from Latin *exempli gratia*, meaning "for example"]

Example: Not all trees are deciduous, <u>e.g.</u>, evergreeen trees never lose their leaves.

    3. Phrase subordinators:

SUCH AS
for example
e.g.

Example: Some elements, <u>such as</u> bromine and mercury, are liquids at room temperature.

Choice: □ or □ (Logical symbol: A∨B)

        Choice words indicate that the idea in the first sentence can be replaced by the idea in the second.

---

    1. Coordinators:

    ON THE OTHER HAND (1,2,3)
    alternatively (1)

Example: The medication can be injected; <u>on the other hand</u>, it can be given intravenously.

    2. Clause subordinators:
    - - - - - - - - - - - - - - - - -
    3. Phrase subordinators:
    - - - - - - - - - - - - - - - - -

Summation: □ □ □ □ → □

        Summation words indicate that the last sentence summarizes the ideas in the previous sentence(s).

---

    1. Coordinators:

    IN CONCLUSION (1,2)   *to conclude* (1)
    in summary (1)   to summarize (1,2)
    in brief (1)   to be brief (1,2)
    in a word (1,2)
    in short (1,2)

Example: The rocket was tested and retested, subjected to all manner of stresses, and flown unmanned into space. <u>In short</u>, all precautions were taken to insure the safety of the astronauts.

    2. Clause subordinators:
    - - - - - - - - - - - - - - - - -
    3. Phrase subordinators:
    - - - - - - - - - - - - - - - - -

           □
Transition: □ ⟶ □

        Transition words indicate (1) that a new aspect of the topic will be discussed or (2) that an unrelated topic will be discussed.

1. Coordinators:

New aspect of topic	Unrelated topic
now (1)	*incidentally* (1,2)
	by the way (1,2)

Examples: [Earlier mention of DNA.] <u>Now in DNA</u>, the replication process pairs op-posite bases.
Fluorine reacts explosively with water; <u>incidentally</u>, so does metallic sodium.

2. Clause subordinators:
- - - - - - - - - - - - - - - - -

3. Phrase subordinators:

New aspect of topic	Unrelated topic
in regard to	- - - - - - - - - - - -
in reference to	
with respect to	

Examples: The report said nothing <u>in regard to</u> the acid-rain problem.
The hospital held a press conference <u>in reference to</u> the successful artifi-cial-heart operation.
<u>With respect to</u> the weather, meteorologists predict mild temperatures.

EXERCISE 2.1   Fill the blanks with coordinators from the AND group be-low. Use synonyms if you want to. Sometimes more than one answer is possible.

in addition	in fact	otherwise	in conclusion
similarly	in other words	first, second	by the way
furthermore	in that case	on the other hand	for example

Example: The fault slips; _____, an earthquake occurs.
 Answer: The fault slips; in that case, an earthquake occurs.

1. Earthquakes cause the land to shake; _____, they sometimes trigger huge waves called tsunamis or tidal waves.
2. Uranium has a half-life of 4.5 billion years; _____, it takes 4.5 billion years for half the atoms in a mass of radioactive uranium to decay to lead (Pb).
3. We followed normal procedures when the patient complained of chest pain. _____, we measured the heart and pulse rate. _____, we administered an electrocardiagram. _____, we had the patient undergo a stress-exercise test.

4. Under some circumstances, light is considered to be a wave; _____, there are conditions under which light behaves as discrete "packets" or quanta.
5. Farmers need more water. Households use water for new appliances and more frequent baths. Underground sources of water are being used up. _____, water reclamation is an absolute necessity.
6. The sun is a very strong source of radio waves; _____, there was a very good program on the radio last night.
7. An iceberg floats on water; _____, the continents float on magma, or liquid rock.
8. In the Jurassic era, dinosaurs lived on land, in swamps and rivers, and in the air; _____, there were even ocean-going types.
9. A cyclone is a violent circular wind that often does great damage; _____, a cyclone struck Andhra Pradesh in India in 1977, killing 10,000 people.
10. Most computers require a maximum operating temperature of 85° F; _____, the chips can overheat, causing storage errors and other problems.
11. Some psychiatric drugs do not help patients at all; _____, some of the powerful tranquilizers are thought to cause serious brain diseases.
12. Some genes suppress the expression of other genes; _____, genes for a particular character may not allow the effects of another gene to show.
13. Some of the different characteristics of an amplifier design are power consumption, balance, input and output resistance, open-circuit voltage amplification, and cost. _____, the design of an amplifier is determined by its intended use.
14. Contrary to popular opinion, ostriches do not eat metal; _____, they do not bury their heads in the sand.
15. A newly discovered infrared object in the solar system could be a burned-out comet; _____, it may be merely an asteroid.
16. Sometimes, the electrical current exceeds the safety limit of a circuit; _____, the fuse blows and the power is disconnected.

EXERCISE 2.2   Choose six of the sentences in Exercise 2.1 and rewrite them with clause subordinators from the AND group below. Remember that not all coordinators have corresponding clause subordinators.

in addition to the fact that	in the same way as	if
not to mention the fact that	by which we mean that	unless

Example: The fault slips; in that case, an earthquake occurs.
  Answer: If the fault slips, an earthquake occurs.

> **EXERCISE 2.3**   Choose four of the sentences in exercise 2.1 and rewrite them with phrase subordinators from the AND group below. Remember that not all phrase subordinators have corresponding coordinators. If you find it impossible to change the sentence into a noun phrase, create a new noun phrase.

> > in addition to   in the event of
> > not to mention   barring

Example: The fault slips; in that case, an earthquake occurs.
  Answer: An earthquake occurs in the event of fault slippage.

THE *BUT* GROUP

Several words show the BUT relationship between two sentences. The different forms of BUT include contrast, concession, reservation, and rebuttal.

Contrast: ☐ but ☐ (logical symbol: A.B)

Contrast words indicate that the second idea is in direct contrast to the first sentence.

---

1. Coordinators:

IN CONTRAST (1,2)
by comparison (1,2)
by way of contrast (1,2)
conversely (1,2)
[on the one hand] on the other hand (1,2)

Examples: Earth is a rocky planet; in contrast, Jupiter consists mostly of gas.
         On the one hand, the body needs salt; on the other hand, too much salt can cause high blood pressure.

2. Clause subordinators:

where
whereas
while

Examples: Earth is a rocky planet whereas Jupiter consists mostly of gas.
        While atomic fission splits an atom, atomic fusion forces atoms together.

---

3. Phrase subordinators:

in contrast with
compared with
by comparison with

Examples: <u>In contrast with</u> Jupiter, earth is a rocky planet.
The damage from the wind was neglible <u>compared with</u> that from last year's storm.

Concession: □ yet □ (Logical symbol: A.B)

Concession words indicate that the second sentence contrasts with the first, but that this information is surprising or unexpected.

1. Coordinators:

HOWEVER (1,2,3)
nevertheless (1,2,3)
nonetheless (1,2,3)
still (1)
in spite of that (1,2,3)

Example: The bomb fell; <u>nevertheless</u>, it did not explode.

2. Clause subordinators:

ALTHOUGH
even though
though [informal]
even if
in spite of the fact that
despite the fact that
notwithstanding the fact that

Example: <u>Although</u> the bomb fell, it did not explode.

3. Phrase subordinators:

in spite of
despite
notwithstanding

Examples: The rocket was launched <u>in spite of</u> the rain.
The patient died <u>notwithstanding</u> his good prognosis.

Reservation: ⟨?⟩ but ☐

      Reservation words indicate that the first sentence may not be completely accurate, but that the second sentence at least is true.

---

1. Coordinators:

AT LEAST (1,2,3)
certainly (1)
that is to say (1,2)

Examples: There are no black holes in the Milky Way; <u>at least</u>, none have been found so far.
The disease was caused by a virus; <u>that is to say</u>, initial studies indicate a viral cause.

2. Clause subordinators:

- - - - - - - - - - - - - - - - - -

3. Phrase subordinators:

- - - - - - - - - - - - - - - - - -

Rebuttal: ☐ but ☐ (Logical symbol: ~A.B)

      Rebuttal words indicate that the first sentence is not a true or correct opinion, but that the second sentence is true or correct.

---

1. Coordinators:

IN FACT (1,2,3)
as a matter of fact (1,2,3)
actually (1)
in reality (1)
instead (1)

Examples: Early navigators thought that the world was flat; <u>in fact</u>, the world is round.
The report stated that the duration was fifty minutes; <u>instead</u>, it was fifty seconds.

2. Clause subordinators:

when in fact

Examples: Early navigators thought that the world was flat <u>when in fact</u> it is round. The report stated that the duration was fifty minutes <u>when in fact</u> it was fifty seconds.

       3. Phrase subordinators:

       instead of

Example: The report stated that the duration was fifty minutes <u>instead of</u> fifty seconds.

EXERCISE 2.4   Fill the blanks with coordinators from the BUT group below. Use synonyms if you want to.

       in contrast   at least
       however   in fact

Example: Earth is a rocky planet; _____, Jupiter consists mostly of gas.
  Answer: Earth is a rocky planet; in contrast, Jupiter consists mostly of gas.

    1. Rabbits have many of the characteristics of rodents; _____, they were placed in a different order (Lagomorpha) because of differences in their teeth.
    2. A cyclotron accelerates protons; _____, a betatron accelerates electrons.
    3. Pulsars were first thought to be signs of intelligent life in outer space; _____, the pulses of radiation come from the star's rotation.
    4. Coffee prevents cancer; _____, an acid found in coffee neutralizes certain cancer-causing pollutants in the body.
    5. Mountains were originally believed to have "roots" supporting them; _____, they are part of the floating crustal plates.
    6. The color white is a mixture of all the colors of light; _____, the color black is the absence of those colors.
    7. Weather forecasting has always been an approximate science; _____, satellites and computers have improved predictions.
    8. Biochemical imbalances can induce criminal behavior; _____, a deficiency of the neurotransmitter serotonin has been linked to aggressive actions, including murder and suicide.
    9. Incandescent lighting is relatively safe; _____, fluorescent lighting may cause a form of skin cancer.
   10. The patient was diagnosed as having a mental disorder; _____, she had a simple thyroid deficiency.

EXERCISE 2.5   Choose five of the sentences from Exercise 2.4 and rewrite them, using clause subordinators from the BUT group below. Remember that not all coordinators have corresponding clause subordinators.

> whereas   although   when in fact

Example: Earth is a rocky planet; in contrast, Jupiter consists mostly of gas.
  Answer: Whereas earth is a rocky planet, Jupiter consists mostly of gas.

EXERCISE 2.6   Choose five of the sentences from Exercise 2.4 and rewrite them, using phrase subordinators from the BUT group below. Remember that not all phrase subordinators have corresponding coordinators. If you find it impossible to change the sentence into a noun phrase, create a new noun phrase.

> in contrast with   in spite of   instead of

Example: Earth is a rocky planet; in contrast, Jupiter consists mostly of gas.
  Answer: In contrast with Jupiter, earth is a rocky planet.

THE *SO* GROUP

The words in the SO group are concerned with cause and effect. These words indicate cause, effect, and purpose.

Cause: □ ← □   (□, for □)   (Logical symbol: A∴B)

Cause words show why something occurs or exists.

---

1. Coordinators:

the reason is [informal] (1)

Examples: The tree died; <u>the reason is</u>, the elephants ate most of the bark.
         The ship sank; <u>the reason is</u>, it struck a large iceberg.

2. Clause subordinators:

BECAUSE
since
as
inasmuch as
due to the fact that

Examples: The tree died <u>because</u> the elephants ate most of the bark.
The ship sank <u>due to the fact that</u> it struck a large iceberg.

     3. Phrase subordinators:

BECAUSE OF
owing to
due to (used only after a linking verb such as *be*)

Examples: The tree died <u>because of</u> the elephants' eating its bark.
The sinking of the ship was <u>due to</u> an iceberg.

Effect: □ → □   (□, so □) (Logical symbol: A∴B)

     Effect words show the result of the previous sentence.

     1. Coordinators:

THEREFORE (1,2)
consequently (1,2)
as a result (1)
thus (1)
accordingly (1,2)
as a consequence (1,2)
for this reason (1,2)

Examples: The elephants ate most of the bark; <u>therefore</u>, the tree died.
The ship struck a large iceberg; <u>as a result</u>, it sank.

     2. Clause subordinators:

so . . . (adjective or adverb) . . . *that*
such . . . (noun phrase) . . . *that*

Examples: The machine got <u>so hot that</u> it melted the wiring.
The elephants ate <u>so much bark that</u> the tree died.
The ship struck <u>such a large iceberg that</u> it sank.
These patients have <u>such high temperatures that</u> their lives are in danger.

     3. Phrase subordinators:
     - - - - - - - - - - - - - - - - - -

Purpose: □ → ⌒⌒

     Purpose words indicate the possibility of a result. Coordinators typi-
cally occur with the verb *want*. Subordinators must occur with a modal, usually

can, will, or *may.* Remember, however, that single events usually require *will,* whereas habitual or regular events (facts) require the simple present tense (see the footnote for condition in this section).

---

1. Coordinators:

for this purpose (1)
for this reason (1,2)

Examples:

Present:
The company wants to save fuel; <u>for this purpose</u>, the output of the steam-power plant has been reduced.
Patients with infectious diseases may infect others; <u>for this reason</u>, they are isolated.
Past:
The patient wanted to sleep; <u>for this reason</u>, the nurse gave him a sleeping pill.

2. Clause subordinators:

so that

Examples:

Present:
The output of the steam-power plant has been reduced <u>so that</u> the company will save fuel.
Patients with infectious diseases are isolated <u>so that</u> they do not infect others.
Past:
The nurse gave the patient a sleeping pill <u>so that</u> he could sleep.

3. Phrase subordinators:
- - - - - - - - - - - - - - - - -

EXERCISE 2.7   Fill the blanks with coordinators from the SO group below. Use synonyms if you want to.

the reason is   therefore   for this reason

Example: The ice melted; _____, the temperature increased.
  Answer: The ice melted; the reason is, the temperature increased.

1. Some animals hibernate during the winter; _____,
   the winter season is too cold to hunt for food.
2. Machines lose power because of friction; _____,
   they must be frequently oiled.
3. Babies with different illnesses have different cries; _____,
   their cries may be used to identify a wide range of later problems.
4. Alkanes consist entirely of single bonds; _____,
   they are very stable and resistant to chemical change.
5. The farmer wanted to kill the weeds in his cornfield; _____,
   he plowed the field and used a selective weed killer.
6. A computer programmer usually makes a flow chart before writing
   a program; _____, a flow chart helps to clarify
   the logical solution to a problem.

EXERCISE 2.8 Fill the blanks with phrase subordinators from the SO group below.

   because of   due to

Example: The ice melted _____ the increase in temperature.
Answer: The ice melted because of the increase in temperature.

1. The leak in the reactor cooling system was _____
   a faulty valve.
2. Many animals migrate to southern climates _____
   extreme winter temperatures.
3. Scientists are aware that the aurora borealis is _____
   the effect of the solar wind on molecules in the earth's atmosphere.
4. The deforestation of certain parts of Africa has been largely
   _____ drought and poor farming practices.
5. _____ his defective immune system, the patient is vulnerable to certain rare diseases.
6. The bridge collapsed _____ insufficient wind
   bracing in the substructure of the roadbed.

EXERCISE 2.9 The following sentences are inversions of the sentences in Exercise 2.7. Fill the blanks with coordinators from the SO group below. Use synonyms if you want to.

   the reason is   for this reason

Example: The temperature increased; _____, the ice melted.
Answer: The temperature incrcased; for this reason, the ice melted.

1. The winter season is too cold to hunt for food; _____,
   some animals have to hibernate.

2. Machines must be frequently oiled; ————————————,
   they lose power from friction.
3. The cries of babies may be used to identify a wide range of later
   problems; ————————————, babies with different illnesses
   have different cries.
4. Alkanes are very stable and resistant to chemical change;
   ————————————, they consist entirely of single bonds.
5. The farmer plowed his cornfield and used a selective weed killer;
   ————————————, he wanted to kill the weeds.
6. A flow chart helps to clarify the logical solution to a problem;
   ————————————, a computer programmer usually makes
   a flow chart before writing a program.

EXERCISE 2.10 Rewrite sentences 1, 2, and 3 in Exercise 2.9 using a clause
subordinator from the SO group below.

> because   so ... that (such ... that)   so that

Example: The temperature increased; therefore, the ice melted.
Answers: The temperature increased so much that the ice melted.
            There was such an increase in temperature that the ice melted.

EXERCISE 2.11 Complete the following purpose sentences using the information given.

Example: The ice must not melt. Therefore, the temperature is decreased.
            The temperature is decreased so that ————————————.
Answer: The temperature is decreased so that the ice will not melt.

1. The demand for oil was not satisfied. Oil tankers were enlarged.
   Oil tankers were enlarged so that ————————————
   ————————————.
2. Expensive equipment was lost. This is the reason that the space
   shuttle was built.
   The space shuttle was built so that ————————————
   ————————————.
3. Many flowers reproduce, thanks to insects.
   Many flowers need insects so that ————————————
   ————————————.
4. Traffic on local freeways is increasing; for this reason, a new
   bridge will be built.
   A new bridge will be built so that ————————————
   ————————————.

5. Milk sours easily. That is why it must be kept in the refrigerator.
   Milk must be kept in the refrigerator so that _____
   _____.

6. Researchers have not yet found a cure for cancer. Thus, they are studying this illness.
   Researchers are studying cancer so that _____
   _____.

7. The silicon chips in a computer must not overheat. For this reason, computers are usually cooled.
   Computers must be cooled so that _____
   _____.

8. The steam valves had to be checked. Therefore, the nuclear-power plant was shut down.
   The nuclear-power plant was shut down so that _____
   _____.

9. Patterns of eruptions are being established; for this purpose, volcanic data are compiled.
   Volcanic data are compiled so that _____
   _____.

10. The dam must not overflow. For this reason, the water level in the reservoir will be reduced.
    The water level in the reservoir will be reduced so that _____
    _____.

THE *TIME* GROUP

The TIME group consists of words and phrases concerned with time (t). Unlike many of the other coordinators and subordinators, they are **not** synonymous. Time words indicate sequence or simultaneity.

Sequence: □ . . . ?t? . . . □ (symbol: $t_A > t_B$; $t_A < t_B$)

Sequence words show that the event in the second sentence took place before or after the event in the first sentence.

---

1. Coordinators:

   a. *before then* (1), *before that* (1), *formerly* (1)
   b. *until then* (1), *until that time* (1), *up to then* (1), *up to that time* (1)
   c. *after that* (1), *afterwards* (1), *later* (1), *subsequently* (1), *then* (1)
   d. *since then* (1), *since that time* (1)
   e. *in time* (1), *eventually* (1)

Examples: a. The silicon chip was invented in 1959; <u>before then</u>, computers were very large.

---

    b. The silicon chip was invented in 1959; <u>until then</u>, computers were very large.

    c. The silicon chip was invented in 1959; <u>after that</u>, computers became much smaller.

    d. The silicon chip was invented in 1959; <u>since then</u>, computers have become much smaller.

    e. The silicon chip was invented in 1959; <u>in time</u>, it will probably be made of a more efficient material.

    2. Clause subordinators:

      a. *before*
      b. *until*
      c. *after*
      d. *since* (used with perfect tenses only)

Examples: a. Computers were large <u>before</u> the silicon chip was invented.
          b. Computers were large <u>until</u> the silicon chip was invented.
          c. Computers became smaller <u>after</u> the silicon chip was invented.
          d. Computers have become smaller <u>since</u> the silicon chip was invented.

    3. Phrase subordinators

      a. *before*
      b. *until*
      c. *after*
      d. *since* (used with perfect tenses only)

Examples: a. Computers were large <u>before</u> the invention of the silicon chip.
          b. Computers were large <u>until</u> the invention of the silicon chip.
          c. Computers became smaller <u>after</u> the invention of the silicon chip.
          d. Computers have become smaller <u>since</u> the invention of the silicon chip.

Simultaneity: ☐ (symbol: $t_A = t_B$)

    Simultaneity words indicate that the event in the first sentence took place at the same time as the event in the second sentence.

    1. Coordinators:

      a. [present] *at this time* (1), *during this time* (1)
          [past] *at that time* (1), *during that time* (1)
      b. *meanwhile* (1,2), *in the meantime* (1,2,3)
      c. *at the same time* (1), *simultaneously* (1)
          [past] *just then* (1)

Examples: a. The earth was young; <u>at that time</u>, the continents were formed.
b. Administer the anesthetic; <u>in the meantime</u>, Dr. Rytz will watch the patient's heart rate.
c. The fuel mixture explodes; <u>simultaneously</u>, the piston moves down.

2. Clause subordinators:

a. *when*
b. *while*
c. *as*

Examples: a. The continents were formed <u>when</u> the earth was young.
b. Watch the patient's heart rate <u>while</u> you administer the anesthetic.
c. The piston moves down <u>as</u> the fuel mixture explodes.

3. Phrase subordinators:

at the time of
during the time of

Examples: No one was in the lab <u>at the time of</u> the explosion.
Flowers appeared <u>during the time of</u> the dinosaurs.

EXERCISE 2.12   Fill the blanks with coordinators from the TIME group below. Use synonyms if you want to.

before that	since then	at that time
until then	at this time	during that time
after that	during this time	at the same time

Example: Newton discovered the law of gravitation in 1687.
_____, he published his findings.
Answer: Newton discovered the law of gravitation in 1687.
After that, he published his findings.

1. Joseph Lister introduced antiseptics in 1865. _____,
   many patients died from infection.
2. First the rock sample is weighed; _____, it is
   crushed and dissolved in acid.
3. The nuclear-engineering staff was not understanding why the
   emergency lights were flashing; _____, the
   core temperature was steadily increasing.
4. First, the root emerges; _____, a shoot appears and begins to turn green.

5. Long ago the Appalachian Mountains were as high as the Alps. _____, they have been gradually worn down by erosion to their present height.
6. A meteor struck the earth 65 million years ago and filled the air with dust particles. _____, a lot of species, including the dinosaurs, became extinct.
7. Bees gather nectar to make honey; _____, they pollinate the flowers.
8. A deep-sea drilling project penetrated 1076 meters into the oceanic crust. _____, 600 meters had been the limit of penetration.
9. From 1840 to 1870, 254,000 emigrants traveled from the Missouri River to the Pacific. _____, nine out of ten deaths were from disease, very few from the Indians.
10. A mechanical arm places a lid on the container; _____, a label is applied, the whole process taking 0.12 second.

EXERCISE 2.13   Choose seven sentences from Exercise 2.12 and rewrite them using clause subordinators from the TIME group below.

before   after   when   as
until   since   while

Example: Newton discovered the law of gravitation in 1687. After that, he published his findings.
Answer: After Newton discovered the law of gravitation in 1687, he published his findings.

EXERCISE 2.14   Choose three sentences from Exercise 2.12 and rewrite them using phrase subordinators from the TIME group below.

before   after   at the time of
until   since   during the time of

Example: Newton discovered the law of gravitation in 1687. After that, he published his findings.
Answer: Newton published his findings after his discovery of the law of gravitation in 1687.

EXERCISE 2.15   Combine the following sentences using any appropriate co-ordinator or subordinator from any of the four groups (AND, BUT, SO, or TIME).

Example: The patient is given an anesthetic.
            The operation is performed.

Possible answers: The patient is given an anesthetic; after that, the operation is performed.

After the patient is given an anesthetic, the operation is performed.

The operation is performed on the patient after the administration of an anesthetic.

1. The continents were forming.
   The moon was only 10,000 miles from the earth.
2. Wild animals with rabies must be killed.
   They might infect other animals or even human beings.
3. Termites are able to digest cellulose in buildings.
   Shipworms can digest cellulose in ships and piers.
4. The satellite fell back to earth.
   It did no harm to human populations.
5. The seismograph indicated a magnitude 7.1 earthquake.
   The volcano began to erupt.
6. Chlorofluorocarbons will reduce the protective ozone layer by 5 to 9 percent in the next decade.
   Humans could stop using them.
7. Ionized nitrogen gas molecules fluoresce as they recombine.
   They emit visible light after an electric current is removed.
8. The glacier passed through this originally V-shaped valley.
   The valley became smooth and U-shaped.
9. Most people believe that Thomas Edison invented the incandescent electric light bulb.
   It was invented by Sir Joseph Swann in England.
10. The power contained in the atom was first demonstrated in 1945.
    The world has seen the development of both military and peaceful applications of nuclear power.

Section 3
REDUCTION AND MODIFICATION

APPOSITIVES

An appositive is a special kind of reduced nondefining relative clause. However, before we discuss appositives, try the following exercise.

EXERCISE 3.1   Remove any unnecessary words in the following sentences.

1. Herpes, which is one of the fastest-spreading diseases in the world, is caused by the herpes simplex virus, type 2.
2. A new high-speed switch chip, which is a 0.2-inch-square piece of silicon, can make connections every 40 billionths of a second.

3. Helium, which is an extremely stable noble gas, was among the atmosphere's earliest components.
4. Meteorites, which are small masses of rock that fall from outer space, were formed about the same time as the earth.
5. Friction, which is a force that resists movement, can be reduced by lubrication but never completely removed.
6. Nature, which is the most important laboratory for natural products, has produced many effective medicines.

An appositive is made by removing the relative pronoun plus *be* from a nondefining relative clause. This reduction can be made **only** when the complement is equal to the antecedent of the relative pronoun. Look at these examples:

Dr. Smith, who is a surgeon at Johns Hopkins, was the first to try the new heart procedure.
Dr. Smith = a new surgeon at Johns Hopkins
                      APPOSITIVE
Dr. Smith, a surgeon at Johns Hopkins, was the first to try the new heart procedure.
Latex, which is the "blood" of the rubber tree, has many industrial uses.
Latex = the "blood" of the rubber tree
                 APPOSITIVE
Latex, the "blood" of the rubber tree, has many industrial uses.

Notice that the appositive can represent a larger group to which the subject belongs (e.g., Dr. Smith is one of many surgeons at Johns Hopkins). For this reason, appositives often serve as implicit definitions, and they are frequently used in scientific text. For example,

"According to the current physical theory, quarks, the objects out of which most subatomic particles are built, should never appear as free bodies."[4]

EXERCISE 3.2 Combine the following sentences into nondefining relative clauses by making the (b) sentence into an appositive phrase.

Example: (a) Pluto is 3675.27 million miles from the sun.
(b) Pluto is the most distant planet in the solar system.
Answer: Pluto, the most distant planet in the solar system, is 3675.27 million miles from the sun.

[4]Reprinted with permission from *Science News*, the weekly newsmagazine of science, copyright 1983 by Science Service, Inc.

1. (a) Mt. Everest is 8848 meters high.
   (b) Mt. Everest is the tallest mountain in the world.
2. (a) The first plastic was developed in England in 1855.
   (b) The first plastic was nitrocellulose softened by vegetable oil and camphor.
3. (a) A ruptured aneurysm often leads to a stroke.
   (b) A ruptured aneurysm is a blood-filled bubble in a vessel.
4. (a) A new hybrid of potato has glue-tipped hairs that trap insects.
   (b) A new hybrid of potato is a cross between wild and cultivated potatoes.
5. (a) The six millionth known chemical is a derivative of a chemical used in the production of pharmaceuticals.
   (b) The six millionth known chemical is 2-cyclohexyl-3-methyl-4(pentylamino)-2-cyclopentene-1-one.
6. (a) One type of cold virus has been crystallized into an icosahedron.
   (b) An icosahedron is a regular polyhedron with twenty sides.
7. (a) Many perfumes, nasal decongestants, and cigarette residues metabolize into formaldehyde.
   (b) Formaldehyde is a known nasal carcinogen.
8. (a) Sleep apnea is much more common among males than females.
   (b) Sleep apnea is a life-threatening disorder characterized by frequently blocked breathing.
9. (a) Carl Wilhelm Scheele actually preceded Joseph Priestly in isolating oxygen.
   (b) Carl Wilhelm Scheele was a Swedish chemist.
10. (a) Lighter fractions of petroleum are preferentially expelled from thin shale layers into overlying sandstones.
    (b) Lighter fractions of petroleum are hydrocarbons with fifteen to nineteen carbons.

Single-Letter Appositives

Single-letter (or sometimes single-number) appositives are commonly used in reference to charts and diagrams. Look at Figure 4.1 on page 168.

Example: In Figure 4.1, water is heated by the Bunsen burner, A, in the distilling flask, B. The steam produced passes into the condenser, C. Distilled water is collected in the beaker, D. In this example,

the Bunsen burner = A
the distilling flask = B
the condenser = C
the beaker = D

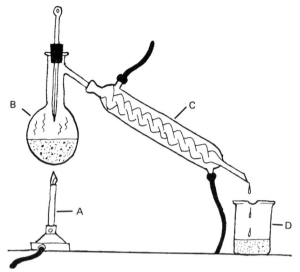

Figure 4.1 Distillation Apparatus

Notice that there is only one example of each different labeled object. If there are **two or more** similar objects, these are often labeled with single-letter names, which are not the same as appositives and do not require commas.

Figure 4.2 shows three different beakers collecting different components of the distillation. These beakers are each named with a letter.

Figure 4.2 Oil Distillation

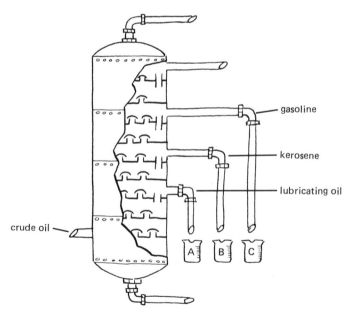

Example: In Figure 4.2, heated oil is introduced into the distillation column. Lubricating oil is collected in beaker A, kerosene is collected in beaker B, and gasoline is collected in beaker C.

Notice that names are different from appositives in that no commas are used and the article *the* is deleted, as is common in names and proper nouns. (See Unit VI, Section 1, "Articles with Proper Nouns"). However, it is also possible to use appositives to express the same information:

Examples: The liquid is delivered into the beaker, D. (appositive).
The liquid is delivered into beaker D. (name)

In Figure 4.2, heated oil is introduced into the distillation column. Lubricating oil is collected in the first beaker, A. Kerosene is collected in the second beaker, B. Gasoline is collected in the third beaker, C.

EXERCISE 3.3   Add the information in parentheses to the blanks.

Example: Extract ——————————— (A = the first molar).
Answer: Extract the first molar, A.
Example: Prepare ——————————— (B = the name of the incisor).
Answer: Prepare incisor B.

1. The voltmeter is connected at ——————————— (D = the name of the plug).
2. The stomach wall consists of ——————————— (p = the peritoneum), ——————————— (s = the serosa), ———— (ml = the muscle layer), ——————————— (sm = the submucosa), and ——————————— (mm = the mucous membrane).
3. Helium II, a superfluid, climbs into ——————————— (1 = the name of the tube) and out of ——————————— (2 = the name of the tube).
4. By measuring the angles at ———— and ———— and the length of line ————, you can determine the distance from the shore ————.

5. In a venturi meter, the loss of charge in a pipe can be determined by comparing the pressure at ——————————— (F = the

name of the upstream pressure tap) to the pressure at _____ (I = the name of the downstream pressure tap).

6. By taking _____ (x = the center of the ring), _____ (y = a point on one ring), and _____ (z = the point where an electron struck a molecule) as three points of a triangle, the operator can determine the angle at which the electron was diffracted.

7. Brake fluid in a hydraulic brake system is forced from _____ (c_1 = the name of the main cylinder) through _____ (bl = the brake line) to _____ (c_2 = the name of the brake cylinder). Consequently, _____ (bs = the brake shoes) are pushed against _____ (bd = the brake drum) and the car comes to a halt.

8.

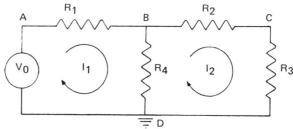

The current flowing from _____ (A = the name of the node) to _____ (B = the name of the node) through _____ (R_1 = the value of the resistance) is equal to _____ (I_1 = the value of the mesh current). The current flowing from _____ (C = the name of the node) to _____ (B = the name of the node) through _____ (R_2 = the value of the resistance) is equal to _____ ($-I_2$ = the value of the mesh current). The current flowing from _____ (B = the name of the node) to _____ (D = the name of the node) through _____ (R_4 = the value of the resistance) is therefore equal to $(I_1 - I_2)$.[5]

OTHER, ANOTHER, THE OTHERS, etc.

Words and phrases containing the word *other* are both postdeterminers (det; see Unit III, Section 1, "The Effect of Modification on Article Choice") and pronouns (pron). In conjunction with *an* and *the,* they indicate either an unlim-

[5] William G. Oldham and Steven E. Schwarz, *An Introduction to Electronics* (New York: Holt, Rinehart & Winston, 1972), pp. 38–39.

ited or a limited relationship to similar nouns. The following chart shows these relationships, using the word *book*.

	UNLIMITED (*an*, Ø)	LIMITED (*the*)
S I N G U L A R	another book (det) another (pron) some/any other book (det) any other (pron)	the other book (det) the other (pron) -
P L U R A L	other books (det) others (pron) some/any other books (det) some/any others (pron)	the other books (det) the others (pron) some of/any of the other books (det) some of/any of the others (pron)

Other and *some other* cannot be used as pronouns.

Definitions and Examples

WORD	DEFINITION	EXAMPLE
another	one more	The engineers found <u>another</u> leak in the dam (i.e., they have already found some leaks and now they have found one more).
	not this one	This terminal is being used. Please use <u>another</u> one.
any other	all except this one	Hydrogen is lighter than <u>any other</u> element (i.e., all elements are heavier than hydrogen). [**Note:** used only in comparisons.]
some other	not this one but one from a defined group	Since the transmission of hepatitis is already being studied, you will have to report on <u>some other</u> aspect of the disease.
other	one more group	These cells were affected; <u>other</u> cells were not.
others	not these	Some people are endomorphs; <u>others</u> are ectomorphs; still <u>others</u> are mesomorphs.
the other	not this one but that one	This eye is infected; <u>the other</u> is all right.
the others (the rest)	not this/these but those	One leaf turned yellow, but <u>the others (the rest)</u> are still green.
any of the others	all the members of a defined group but not this/these	The first patient had a stronger reaction than <u>any of the others</u> in the study. [**Note:** used only in comparisons.]
some of the others	not these but a few from a defined group	Two rats died from radiation exposure. <u>Some of the others</u> became ill and recovered. The rest were completely unaffected.

The rest has the same meaning as *the others* but emphasizes "all that remain."

EXERCISE 3.4 Fill the blanks with one of the words or phrases from the list below.

another (det)	others (pron)
another (pron)	some other (det)
any other (det)	some others (pron)
any other (pron)	the other (det)
any others (pron)	the other (pron)
other (det)	the others (pron)

Example: Many people hear better with one ear than _____.
Answer: Many people hear better with one ear than the other.

1. The laboratory published three reports. The first was accepted; _____ were rejected immediately.
2. One of the group of elements called the halogens is chlorine; _____ is iodine.
3. The Atlantic is one of the two largest oceans on the earth; _____ is the Pacific.
4. Many of the stars in the night sky are within our own galaxy. _____ stars are in neighboring galaxies. Still _____ represent whole galaxies themselves.
5. There is more acid rain today than at _____ time in history.
6. The left and right hemispheres of the brain are not lateralized until after puberty. Consequently, if one side is damaged before this time, _____ side can take over the same functions.
7. If a substance devoid of known impurities does not melt at the proper temperature, a researcher knows that _____ impurity must be in the sample.
8. A lizard can regenerate its limbs. If a leg or the tail is lost, the lizard can grow _____ one.
9. Some cancer patients experience unbearable pain while _____ patients experience only weakness.
10. Electrons can be excited to a higher p-level, but they actually prefer the most stable state to _____.
11. Since an eclipse of the sun will occur precisely at that hour, the team will have to choose _____ day to measure the sun's output.
12. The influenza inoculation was administered only to those patients who were here at eight o'clock sharp. _____ will have to come back next week at the same time.
13. These test tubes are dirty. There are _____ in that cabinet. Could you get them for me?

14. Two of the planets in the solar system are gas giants; _____ _____ are all rocky planets.
15. Two resistors are already in place, as you can see in the schematic diagram. Connect _____ resistor across the circuit and the output radically decreases.

PREPOSITIONS OF TIME

In Unit III, we saw that *at* indicates a specific time, *on* indicates a day or date, and *in* indicates a month, season, year, decade, century, etc. Now we shall discuss the prepositions that indicate (1) range, (2) starting point, (3) end point or limit, and (4) time period.

Range: *from, to, through, between, and until* ○ ⟷ ○

These prepositions indicate a range of time that (1) begins with the object of *from* and ends with the object of *to, through,* or *until,* or (2) is described by *between* and *and.* Look at these examples:

The comet was visible from   March to   May.
The comet was visible from   March until   May.
The comet was visible from   March through May.
The comet was visible between March and   May.

In these sentences, *to, until,* and *between* suggest "up to the beginning of May," whereas *through* means "up to the end of May."

Until indicates that at the described time, the opposite action took place; in other words, in May the comet became invisible. The idea of opposite action taking place is important because, unlike *to, through,* or *between, until* can be used with a negative verb to indicate a period of time before something begins. With a negative verb, *until* indicates not a range but a starting point (see the next section). Look at the following diagrams:

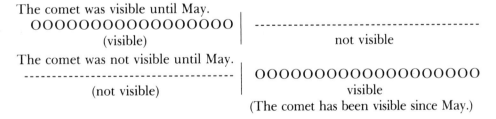

The comet was visible until May.
OOOOOOOOOOOOOOOOOO | --
(visible) | not visible

The comet was not visible until May.
-- | OOOOOOOOOOOOOOOOOOOO
(not visible) | visible
                                   (The comet has been visible since May.)

Until also functions as a clause subordinator (see "The TIME Group" in Section 2 of this Unit):

The volcano did not erupt until there was enough magma in the magma chamber.

The volcano erupted until all the magma in the magma chamber was expelled.

EXERCISE 3.5   Fill the blanks with *from, to, through, until,* or *between . . . and.*

Example: Dr. Fong is on duty _____ 10 P.M. _____ 6 A.M. tonight.
Answer: Dr. Fong is on duty from 10 P.M. to 6 A.M. tonight.

1. The Age of Mammals extends _____ 60 million years ago _____ the present.
2. The planet was visible _____ May _____ the whole month of June.
3. The reaction continued _____ all the aluminum was consumed.
4. The cyclotron is open to the public _____ 10 A.M. _____ 4 P.M. on weekdays.
5. The patient's chance of recovery was very good _____ she got pneumonia.
6. The El Niño current, a massive ocean-atmosphere disturbance, had its greatest impact _____ spring 1982 _____ July 1983.
7. The greatest snowfall ever recorded, 193 cm in twenty-four hours, occurred in Colorado _____ April 14 _____ April 15, 1921.
8. The experiment will begin in March and continue _____ the Arctic summer _____ the first snowfall.
9. The average life span of a cat is _____ ten _____ twelve years.
10. The sun's output gradually declined _____ February 1980 _____ August 1981, then began to rise again.

Starting Point: *since* and *for* ○ →

Both *since* and *for* indicate the time at which something begins. The preposition *since* must be used with a perfect tense. The perfect tenses always require a reference time (see "The Present Perfect Tense" in Section 4, of this unit). In the present perfect, the reference time is NOW. In the past perfect, the reference time is a SIMPLE–PAST VERB. In the future perfect, the reference time is a SIMPLE–PRESENT VERB. Look at the following diagrams:

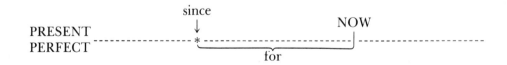

Example: The sun <u>has existed</u> for 4.5 billion years.

PAST
PERFECT

Example: Steam <u>had existed</u> for many years before Thomas Savery
APPLIED it to an engine.

FUTURE
PERFECT

Example: The space probe <u>will have traveled</u> for sixteen years by the time it
REACHES Pluto.

Since indicates the name of a time (e.g., *Saturday, 1976, 3:00* P.M., *his childhood*). *For* indicates the period between the "name of the time" and the reference time (e.g., *four days, ten years, several hours, his entire life*). Look at the following present-perfect examples:

Mr. Smith has been in the hospital . . .

SINCE	FOR
since Saturday.	for four days.
since 1976.	for ten years.
since 3:00 P.M.	for several hours.
since his childhood.	for his entire life.

Unlike *since, for* is not restricted to the perfect tenses:

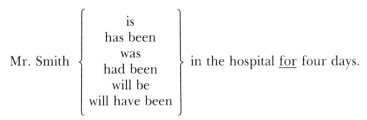

Mr. Smith { is / has been / was / had been / will be / will have been } in the hospital <u>for</u> four days.

In addition, *for* is often deleted in informal writing:

Mr. Smith has been in the hospital four days.

EXERCISE 3.6   Fill the blanks with *since* or *for*.

Example: Airplanes have crossed the Atlantic ———— 1919.
 Answer: Airplanes have crossed the Atlantic since 1919.

1. The antibiotic penicillin has been available ———— 1938.
2. Life on earth has existed ———— approximately 3.5 billion years.
3. The atomic age has existed ———— the first atom-bomb test on July 16, 1945, at 5:30 A.M.
4. The hydrogen fuel supply of the sun will last ———— another 5 billion years.
5. The elm-tree population of the world has greatly diminished ———— the appearance of the Dutch elm disease.
6. The big-bang theory states that the universe has existed ———— its explosive beginning 15 billion years ago.
7. The disintegration of uranium 235 continues ———— billions of years.
8. Mount St. Helens, which became active in 1980, had not erupted ———— 127 years.
9. The patient's heart rate has been abnormal ———— yesterday.
10. The Age of Reptiles lasted ———— 135 million years.

End point/Limit: *by* and *(with)in* ⟶ ○

     The prepositions *by* and *within*, in most cases shortened to *in*, represent a limit or the point at which something ends. *By* is similar to *since* in that it indicates the "name of the time" (e.g., *Saturday, 1976, 3:00 p.m., the time he retires*). *In* is similar to *for* in that it indicates the period between the moment of speaking and the "name of the time." *In* can occur with all tenses. *By* can occur with all tenses except the present perfect tense.

TIME	BEGINNING	END
"NAME"	since	by
PERIOD	for	(with)in

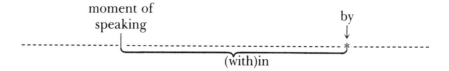

    Look at these examples with *by* and *(with)in:*

The machine must be finished:

BY	(WITH)IN
by Saturday.	in four days.
by 1996.	in ten years.
by 3:00 P.M.	in a few hours.
by the time he retires.	in his working lifetime.

(With)In does not usually occur in negative sentences.

Incorrect:
*The dam, which was supposed to be completed by now, will not open <u>in</u> three years.

In this case, *for* must be used:

Correct:
The dam, which was supposed to be completed by now, will not open <u>for</u> three years.

The expressions *in the morning, in the afternoon, in the evening,* and *in time* come from the preposition *within,* not *in.* For example, *in the morning* can be understood to mean "within the period that we call morning."
The only case where *within* **must** be used is in the phrase "X occurs within [length of time] of Y":

The ambulance arrived <u>within a few seconds of</u> the doctor.

EXERCISE 3.7   Fill the blanks with *by* or *(with)in*.

Example: The X ray will be ready _____ ten minutes.
 Answer: The X ray will be ready in ten minutes.

1. The Golden Gate Bridge was completed _____ 1937.
2. The explosion takes place _____ a fraction of a second.
3. Some depressed patients feel better _____ the daytime.
4. The woolly mammoth was extinct _____ 10,000 B.C.
5. _____ the time the reactor was brought under control, most of the nuclear fuel was uncovered.
6. Organs must be transplanted _____ forty-eight hours of removal.
7. All the fruit must be harvested _____ the first frost.
8. The ambulance rushed the unconscious man to the hospital _____ less than eight minutes.
9. The final plans for the building are due _____ 8 A.M. Monday morning.

10. The medicine should be taken once _____ the morning and once _____ the evening.

Time Period: *during, throughout,* and *over*

The preposition *during* indicates "at some point in a certain time period." Look at these examples:

The patient fell asleep <u>during her examination</u>.
The booster rocket is ejected <u>during the first phase</u>.

During is rarely used with perfect tenses, and the object of *during* must be an event (e.g., *the examination, the first phase, the war, my vacation, the class, [the month of] April, [the year of] 1985*), not the duration of that event in units of time. Duration in units of time is shown by the preposition *for*:

Incorrect:
*John studied there <u>during six years</u>.
Correct:
John studied there <u>for six years</u>.

The preposition *throughout* indicates "for the whole time." Look at these examples:

It rains here <u>throughout the year</u>.
The patient was conscious <u>throughout the operation</u>.

Unlike *during, throughout* often occurs with perfect tenses, but like *during,* the object of *throughout* must be a complete unit of time, not duration in units of time.

Incorrect:
*Mary was an excellent chemist <u>throughout twenty-five years</u>.
Correct:
Mary was an excellent chemist <u>throughout her twenty-five-year career</u>.
Mary was an excellent chemist <u>for twenty-five years</u>.

The time preposition *over* indicates a repeated or continuous action leading to a particular result.

The action of wind and water <u>over time</u> can change the shape of rocks and land contours.

EXERCISE 3.8 Fill the blanks with *during, throughout,* or *over.*

Example: The building collapsed _____ the earthquake.
Answer: The building collapsed during the earthquake.

1. Wolves stay with one mate _____ their lifetimes.
2. Bridalveil Falls in Yosemite National Park was formed _____ the last ice age.
3. The volcano continued to erupt intermittently _____ the next several weeks.
4. No one will be allowed to enter the studio _____ the recording session.
5. Some tumors develop _____ periods of several years.
6. Ice can develop on the wings of planes _____ cold weather.
7. Radioactive-exposure badges must be worn _____ the test procedure.
8. Our planet was formed _____ millions of years from the same primordial material as the sun.
9. Plants require sunlight _____ their growing periods.
10. The mayfly mates once _____ its brief existence.

EXERCISE 3.9 Fill the blanks with the appropriate preposition of time from the list below.

from	until	for	during
to	between . . . and	by	throughout
through	since	(with)in	over

Example: The comet was visible _____ midnight _____ 4 A.M.
Answer: The comet was visible from midnight to 4 A.M.

1. The Sahara did not become a desert _____ the end of the last ice age.
2. The field should be plowed and ready for planting _____ early May.
3. This patient has had a pain in his lower back _____ an automobile accident several years ago.
4. Products are checked _____ the production process, not just when they are finished.
5. The hurricane hit coastal areas with high winds and waves _____ thirty-six hours or more.
6. Thunder is usually heard _____ a few seconds of a lightning flash.
7. The gestation period of an elephant is _____ 510 _____ 730 days.

8. The land was largely lifeless ———— the Lower Cambrian ———— the Upper Devonian subperiods of the Paleozoic era.
9. A single lava sample is insufficient; samples must be taken ———— a period of several days.
10. Stars radiate enormous energy ———— their formation.

Section 4
VERBS

The Present Perfect Tense

The present perfect tense describes the effect of a past action on a present condition (hence the name **present** perfect). It can be described as a past image projected onto the present. This emphasizes the **effect** of the past action rather than the action itself, as depicted in the diagram below.

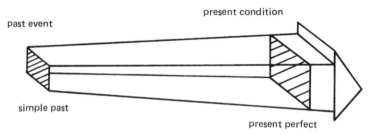

Examples:

The Nobel Prize <u>has been awarded</u> to over 500 scientists.
(I.e., it is recognized as a great honor and will probably continue to be recognized as such.)
The computer <u>has been</u> down twice this week.
(I.e., the system is not perfect and we hope to fix it soon.)
The biology department <u>has assembled</u> over 10,000 specimens.
(I.e., it is a valuable department that works hard.)

Present Perfect versus Simple Past

Since the present perfect emphasizes the effect of a past action and not the time of that action, it is not used with a specific time phrase unless it is necessary to indicate a starting point (see "Starting Point," Section 3 of this unit). Specific time phrases usually require the simple past tense.

Simple Past:
The company <u>built</u> the dam five years ago.
Present Perfect:
The company <u>has built</u> several dams.

The first sentence emphasizes the action of building. The second sentence emphasizes the attributes of the company: it has a lot of experience building dams. As you can see, the implications of the simple past tense and the present perfect are quite different. Here is another example:

Simple Past:
The lab <u>tested</u> the components of the machine.

Implication:
Testing occurred. Logical questions: When were the components tested? How were they tested?

Present Perfect:
The lab <u>has tested</u> the components of the machine.

Implication:
The machine can now be assembled. Result: a dependable machine.

It is typical for a speaker to first ask an unfamiliar person a question in the present perfect tense in order to determine the extent of his or her experience or knowledge. (This is similar to the first-mention rule for articles. See Unit II, Section 1.)

<u>Have you (ever) worked</u> with radioactive materials?
(I.e., Do you have this experience?)

If the answer is *no,* no further questions need be asked. If the answer is *yes,* further questions are usually asked in the simple past. (This is analogous to the second-mention rule for articles.)

When did you work with radioactive materials?
Where did you work?
How long did you work?
With whom did you work? etc.

EXERCISE 4.1   Put the correct present perfect, simple past, or simple present active or passive verb form in the blanks.

Example: Winds and water (erode) _____ the ocean cliffs.
 Answer: Winds and water erode the ocean cliffs.

1. The role of blood clots in heart attack (not yet determine) _____
_____ . But since a clot (block) _____
blood flow to a part of the heart muscle, thereby killing it, researchers hope that removing a clot within the first hour will (resupply) _____ the area with blood.[6]

[6]Reprinted with permission from *Science News,* the weekly newsmagazine of science, copyright 1983 by Science Service, Inc.

2. The outer surfaces of large meteorites (melt) _____ rapidly. After passing air streams (sweep) _____ this liquid rock away, a new cool layer of material (expose) _____ _____ . The molten liquid (create) _____ the blazing trail which (see) _____ from the earth.[7]

3. The earth's internal forces of uplift are more active in the western U.S. than in the eastern part. Eruptions (occur) _____ in this area for centuries. Mount St. Helens (erupt) _____ in 1980. The Baldwin Hills area of the Los Angeles Plain (steadily push) _____ upward at the rate of three feet every 100 years. There (be) _____ also constant slipping along the San Andreas fault.[8]

4. Although man (climb) _____ the highest mountain and (probe) _____ the depths of the ocean, no one (find) _____ sufficient evidence to identify the famous Loch Ness monster in Scotland. However, some researchers (believe) _____ that the "monster" (be) _____ simply pine logs which (fall) _____ into the deep lakes and (decay) _____ , producing gas bubbles which (seal) _____ _____ by the pine resin. Eventually, the logs (become) _____ buoyant, and they (float) _____ to the surface.[9]

5. Not only numbers but all characters (code) _____ in binary when they (transmit) _____ to the computer. The machine (give) _____ a list of equivalent expressions to represent these symbols. The central processing unit (design) _____ to recognize these binary numbers. The programmer, however, must (organize) _____ and (separate) _____ the information the computer needs and the instructions it must (execute) _____ in order to obtain accurate results for the program.[10]

6. Paleontologists (discover) _____ what they (believe) _____ to be the oldest form of life on earth. Life, in the form of bacterial cells, (exist) _____ 3.5 billion years ago, only one billion years after the earth (form) _____ . The fossilized bacteria (discover) _____ by examining old sedimentary rocks (find) _____

[7]Adapted from "Meteorites," in *Compton's Pictured Encyclopedia*, © 1964 by Encyclopaedia Britannica, Inc.

[8]Adapted from Jane Werner Watson, *The World We Live In* (London: Wm. Collins Sons, 1957), p. 53.

[9]Adapted from Robert P. Craig, "Explaining 'Nessie,' " *Newscientist*, August 5, 1982.

[10]John C. Keegel, *The Language of Computer Programming* (New York: Regents Pub. Co., 1976), p. 47.

in Australia. About five different forms of bacteria (identify) _____ that are almost identical to their modern-day cousins. The discovery (push) _____ back the origin of life 1.2 billion years earlier than (previously think) _____ .[11]

7. When a solution of dichromate and sulphuric acid (drip) _____ into boiling ethyl alcohol, acetaldehyde (form) _____ in a medium whose temperature is 60 degrees above its boiling point. For this reason, the acetaldehyde (escape) _____ before the greater part of it (undergo) _____ oxidation. The reaction (carry out) _____ under a fractionating column that (allow) _____ the acetaldehyde to pass but (return) _____ the alcohol to the reaction vessel.[12]

VERBS WITH INANIMATE SUBJECTS

Scientific English is characteristically direct and concise. We have discussed several forms of reduction (see Units II and III, Section 3) for the deletion of repeated words. Now we will see how certain verb structures work in the same way. Look at the following examples based on Figure 4.3.

Figure 4.3 Adding Hydrochloric Acid to Ammonium Chromate

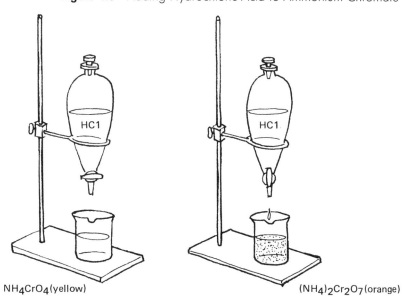

NH$_4$CrO$_4$(yellow)   (NH$_4$)$_2$Cr$_2$O$_7$(orange)

[11]From *Information Please Almanac 1981*. Copyright © 1980 by Houghton Mifflin Company. Reprinted by permission of Houghton Mifflin Company.
[12]Robert Thornton Morrison and Robert Neilson Boyd, *Organic Chemistry*, 2nd Ed. (Boston: Allyn & Bacon, 1966), p. 625.

1. The color of the solution is changed by adding acid.
2. With acid, the color of the solution changes.
3. We change the color of the solution by means of acid.
4. If we add acid, the color of the solution changes.
5. Because acid is present, the color of the solution is changed.

These are all correct sentences in English, but in scientific writing they are considered wordy (see Unit V, Section 5, "Wordiness") and indirect. A better way to write the above sentences is:

Acid changes the color of the solution.

Notice that the subject of the sentence *(acid)* is not a person but an agent (see Unit II, Section 4). In many languages, it is impossible to use an inanimate subject in this way. However, scientific English prefers the use of inanimate subjects and active verbs for clarity and conciseness. Inanimate subjects show or imply a cause-and-effect relationship. Look at these examples:

	Cause	Effect
Gravity keeps the moon in its orbit.	gravity	moon in orbit
The carbon 14 dating method determined the age of the sample.	C 14 method	age of sample determined
The clouds are moving to the east.	(wind, earth rotation)	clouds move
Floods have ruined the crops.	floods	ruined crops

EXERCISE 4.2   Fill the blanks with an inanimate subject and an active verb derived from the information in the first sentence.

Example: In orange juice, vitamin C is contained.
 Answer: Orange juice contains vitamin C.

1. Because of the El Niño current, many fish were killed.
   _____ many fish.
2. In chlorophyll, magnesium is contained.
   _____ magnesium.
3. With the big-bang theory, the origin of the universe is described.
   _____ the origin of the universe.
4. The patient's heartbeat was monitored by the machine.
   _____ the patient's heartbeat.
5. A gas-pipeline connection is made between the oil fields and the seaport.
   _____ the oil fields and the seaport.

6. With nucleophilic addition, aldehydes can be reduced to primary alcohols.
   _____ aldehydes to primary alcohols.
7. Electrical engineers measure the voltage in a system with a voltmeter.
   _____ the voltage in a system.
8. The I-beam has a resistance to compression and tension.
   _____ compression and tension.
9. The heat of the sun causes ocean water to evaporate, forming clouds.
   _____ ocean water, forming clouds.
10. The chain reaction in a nuclear-power plant is shown in Figure 1.
    _____ the chain reaction in a nuclear-power plant.

EXERCISE 4.3   Change the following causes and effects into sentences containing inanimate subjects and active verbs.

Example: catalyst is present → reaction rate is increased
 Answer: The catalyst increases the reaction rate.

1. glaciers existed → many valleys were changed from a V shape to a U shape
2. penicillin is effective → dangerous bacteria are destroyed
3. a seismometer is a machine → it is used to measure earthquakes
4. a meteorite was present → the Arizona desert was struck 30,000 years ago
5. Hurricane Fifi was formed → 8000 Hondurans were killed in 1974
6. welding is used → sometimes metallurgical changes in materials are produced
7. nitrogen is deficient → leaves become yellowed and plant growth is stunted
8. use fire to polish the sharp edges of a new glass tube (Fire polishing . .) → the edges are softened by this process
9. a solution is hypertonic → blood-cell shrinkage occurs
10. a force acts on a rigid body → a reaction force in the opposite direction is generated

EXERCISE 4.4   Underline the inanimate subjects and active verbs contained in Exercise 4.1.

Example: Winds and water erode the ocean cliffs.
 Answer: Winds and water erode the ocean cliffs.

Section 5
WRITING AIDS

TABLES AND FIGURES

Tables and figures are used often in scientific writing because they present a large amount of information in a small space.

Tables

A table usually allows a reader to compare statistical information (see Table 4.1). The following rules should be followed when using a table in scientific writing:

1. The table must have a title.
2. The table **must** be referred to in the text so that the reader knows when to look at it (see Unit III, Section 5, "Referring to Sequential Diagrams").
3. The table should be placed shortly **after** it is first mentioned in the text.
4. One table should fit onto one page.
5. A table from an outside source must be acknowledged like any other borrowed information.

TABLE 4.1 WATER SUPPLY OF THE WORLD

TYPE OF WATER	SURFACE AREA (SQUARE MILES)	VOLUME (CUBIC MILES)	PERCENTAGE OF TOTAL
Salt Water			
Oceans	139,500,000	317,000,000	97.2
Inland seas	270,000	25,000	0.008
Fresh Water			
Lakes	330,000	30,000	0.009
Rivers	—	300	0.0001
Antarctic Icecap	6,000,000	6,300,000	1.9
Arctic Icecap	900,000	680,000	0.21
Atmosphere	197,000,000	3,100	0.001
Ground water	—	2,000,000	0.62
TOTAL		326,000,000	100.00

Source: U.S. Geological Survey, 1980.

Figures

Figures include bar charts, graphs and curves, pie diagrams, organization and flow charts, photographs, and diagrams (see Figure 4.4). In scientific writing, figures are always functional, never decorative. The following rules should be followed when using a figure:

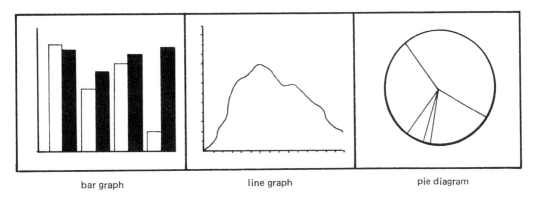

bar graph   line graph   pie diagram

Figure 4.4 Examples of a bar graph, a line graph, and a pie diagram.

1. The figure must have a number and a title or caption.
2. The figure may also have a legend, a short explanation directly below the figure.
3. The figure **must** be referred to in the text so that the reader knows when to look at it (see Unit III, Section 5, "Referring to Sequential Diagrams").
4. The figure should be placed shortly **after** it is first mentioned in the text.
5. One figure should fit onto one page.
6. A figure from an outside source must be acknowledged like any other borrowed information.

PART II:
Writing a Classification

Section 6
PREWRITING ACTIVITY:DETERMINING CLASSIFICATIONS

A classification is used in scientific writing to describe groups of similar objects, substances, and ideas. In order to adequately classify a group of objects, the writer must develop a good system of classification. The following exercise will help you to do this.

EXERCISE 6.1   With another student, choose one of the topics from the list below. (1) On a piece of paper, list all the examples that you can think of for this topic. Compare your list with that of your partner. (2) With your partner, classify all your examples into three or four logical, parallel groups. Give each group a name. (3) Delete those groups that are not parallel and those that include examples that could fit into another group.

Example: types of food

		Types		
beef	cookies	cheese	yogurt	nuts
fish	french fries	tomatoes	eggs	oranges
lettuce	ice cream	bread	cereal	milk

		Groups	
Protein	Dairy products	Fruits/Vegetables	Grains
beef	cheese	lettuce	bread
fish	yogurt	french fries	cereal
eggs	milk	oranges	cookies
nuts	ice cream	tomatoes	

1. kinds of pollution	7. kinds of diseases
2. applications of computers	8. kinds of plants
3. kinds of clothing	9. job classifications
4. types of measuring instruments	10. kinds of transportation
5. kinds of catalysts	11. kinds of structural materials
6. types of metals	12. types of geological formations

Section 7
STRUCTURE

A typical classification introduces the subject, describes each classified group and its members, and makes a conclusion. If a subject has three categories, the classification usually consists of five paragraphs: one for the introduction, one for each category, and one for the conclusion. Of course, more paragraphs would be required if a lot of examples are included in each category. Very long paragraphs are difficult to read.

PARAGRAPH 1: INTRODUCTION

The introduction to a classification consists of three to four sentences in the following order:

1. Formal definition: What is the subject being classified?
2. Purpose: What is the purpose of the subject?
3. Value of classification (optional): Why have the group members (species) been classified in this way and not in some other way?
4. Plan-of-development sentence: What are the categories that will be described?

Formal Definition

A formal definition (see Unit I, Section 2) is one way to begin a classification. However, the definition is sometimes included in a general discussion of the subject rather than being the opening sentence.

Purpose

If the purpose of the subject is not expressed in the formal definition, it can be expressed in a second sentence, which often implies the importance of the subject in some manner.

Example: food
The purpose of food is to provide the body with energy through oxidation.

Value of Classification (Optional)

Different classifications of a subject serve different purposes. For example, food can be classified according to (a) the type of nutrients it contains, (b) the number of calories it contains, (c) the types produced in different areas of the world, or (d) the types of chemicals or minerals it contains. This sentence describes the basis of the classification that has been selected.

Example: food
A well-balanced diet depends on food from a variety of sources.
Foods can be classified according to their caloric content.
Tropical areas produce different foods than do temperate ones.
Different foods contain different minerals.

Plan-of-development Sentence

The plan-of-development sentence tells the reader the order of presentation of the categories in the classification. There are several ways to write the plan-of-development sentence in a classification:

X can be classified $\left\{ \begin{array}{l} \text{as} \\ \text{into} \end{array} \right\}$ A, B, and C.
X can be divided into the following groups: A, B, and C.
X is classified as A, B, and C.

Example: food
Food can be classified into carbohydrates, fats, and proteins.
Food can be divided into the following groups: carbohydrates, fats, and proteins.
Food is classified as carbohydrate, fat, and protein.

There are certain rules that must be followed in setting up the categories in a classification. The categories:

1. must be parallel (i.e., each category must have equal weight).
2. must be independent.
3. must not overlap (i.e., have members from another category).
4. must be complete (i.e., no items left out).
5. must be clear, useful, and purposeful.
6. can be ordered by time, importance, complexity, or location/position.

EXERCISE 7.1   Look at the categories you determined in Exercise 6.1. Form them into a plan-of-development sentence using one of the model sentences above.

Example: Food can be classified into carbohydrates, fats, and proteins.

PARAGRAPH 2: DESCRIPTION OF CATEGORY A

The second paragraph in a classification describes the first category (genus) in the plan-of-development sentence. The first sentence identifies the category (What is it?). The second sentence lists the species. The third and subsequent sentences describe the species in detail. Alternatively, the second sentence may directly identify and describe the first species, the third sentence may identify and describe the second species, and so on. In a classification, the body (paragraphs 2, 3, and 4) is the longest part of the paper. The introduction and conclusion are usually fairly short.

> Example: food
> Carbohydrates are compounds containing carbon, hydrogen, and oxygen. The carbohydrates we eat are sugars and starches. Most sugars have two or more molecules of glucose, fructose, or both joined in one molecule. The starch molecule has hundreds of simple sugar units. Plants also produce a carbohydrate called cellulose, but this is not digestible by humans.
> Carbohydrates are compounds containing carbon, hydrogen, and oxygen. Sugars usually have two or more molecules of glucose, fructose, or both joined in one molecule. Starches have hundreds of simple sugar units. Cellulose is the only carbohydrate not digestible by human beings.

PARAGRAPH 3: DESCRIPTION OF CATEGORY B

Paragraph 3 is constructed in the same manner as paragraph 2.

PARAGRAPH 4: DESCRIPTION OF CATEGORY C

Paragraph 4 is constructed in the same manner as paragraph 2.

PARAGRAPH 5: CONCLUSION

The most common ways to write a conclusion for a classification are as follows:

1. Describe the advantages of either the subject or the classification.
2. Summarize the main points.
3. Describe new ways of classifying the subject, perhaps based on new findings or techniques. For example, food may be classified according to its carcinogenic propensities, its light-scattering patterns, or the possibility of its cultivation in outer space.

There are also other ways to present classifications. Some writers prefer to use a sentence-outline format, which shows the categories very clearly and does not require a plan-of-development sentence (see Unit III, Section 7,

"Paragraph 5: Conclusion"). Another way to present a classification is first to describe broad categories of classification, then to divide the broad categories into two or three subgroups, and finally to describe the species in each subgroup.

Section 8
MODELS

MODEL 1: STATIONARY BRIDGES[13]

Stationary bridges are structures that allow people, motor vehicles, and trains to cross rivers, valleys, and other natural obstacles. A stationary bridge is sufficiently high to permit ships or vehicles to pass under it. Stationary bridges may be classified according to their basis of support. There are three major types: beam bridges, arch bridges, and suspension bridges.

A beam bridge consists of a horizontal member resting on a vertical support at each end. The four most common types include simple beam, continuous beam, truss, and cantilever bridges. A simple beam bridge is like a log across a stream. A modern example is the plate girder bridge, commonly built over highways. A plate girder is a built-up beam consisting of a steel plate to which angles are riveted or welded. A continuous beam bridge is one which resists tension and compression by resting on more than one support.

The truss bridge is made up of members forming rigid triangles. Members in compression are built as rectangles to resist buckling whereas those subject to tension are usually slender bars. A cantilever bridge utilizes a beam that extends beyond its support. Sometimes two cantilever arms meet at mid-span; however, they usually do not meet but are connected by a light, suspended span.

Unlike the beam bridge, in which load is transmitted vertically to the supports, an arch bridge pushes outward against its supports. These must be heavy to resist the horizontal thrust of the arch. Arch bridges differ according to the way in which they carry the horizontal roadway. The four most common types are the deck arch, the half-through arch, the through arch, and the rigid frame arch bridge. In the deck arch, the road is carried above the arch which supports it. In the half-through arch, the road cuts through the arch. Its middle section is hung from the arch, and its outer sections are supported by the arch. In the through arch, the road is suspended from the arch.

The rigid frame is an arch bridge of rectangular shape. It is used for shorter spans, such as in highway and railway grade separations. Its horizontal span is made in one piece with the vertical supports. The rigid frame is made of steel or reinforced or prestressed concrete.

Suspension bridges derive their support from huge cables which are hung over two high towers. As in the arch bridge, the thrust on the suspension

[13]Adapted from "Bridges," in *Compton's Pictured Encyclopedia* © 1964 by Encyclopaedia Britannica, Inc.

bridge is horizontal through the suspender cables which hang from the main cables. Instead of horizontal compression, however, there is horizontal tension on the cable anchorages. Cables are parallel wire, wire rope strand, or eyebar. The parallel wire cable is best for long spans like the Golden Gate Bridge in San Francisco. Thousands of wires are laid side by side, then squeezed together and wrapped with wire to protect them. The wire rope strand consists of prestressed twisted wire strands. Its great advantage is that it is ready for erection without spinning. An eyebar is a piece of steel with a circular head at each end in which there is a hole. In an eyebar bridge, the eyebars are connected with pins, as in the Florianopolis bridge in Brazil.

The greatest triumph of bridge building is the suspension bridge, and still longer suspension bridges are planned, such as a 5000-foot span between Italy and Sicily. Engineers believe that even 10,000-foot spans are feasible.

MODEL 2: ANESTHETICS: PAIN-KILLING DRUGS[14]

Anesthetics are drugs causing unconsciousness or insensibility to pain. Their use in modern medicine permits painless surgery during the simplest operation of a few minutes' duration to the most delicate operation lasting many hours.

Anesthetics are divided into two broad groups: general anesthetics and local anesthetics. General anesthetics cause total unconsciousness in the patient by temporarily altering the normal activities of the central nervous system. Local anesthetics temporarily deaden sensation in a particular, or local, area of the body.

General anesthetics are usually administered to the patient in one of two ways: inhalation or intravenous injection. In the inhalation method the patient breathes a gas or vapor into his lungs. In the intravenous injection method the drug is put directly into a vein.

Several important drugs are used as general anesthetics. Ether, first used by Dr. Crawford W. Long in 1842 to anesthetize a patient, is a powerful anesthetic. As a powerful liquid that is inhaled, it can be irritating to respiratory membranes. Another potent general anesthetic is cyclopropane. This gas is inhaled in combination with oxygen. A much weaker general anesthetic is nitrous oxide, or laughing gas, which is usually combined with other drugs. Ethylene is a slightly stronger gas than nitrous oxide and has an unpleasant odor. Chloroform is a very powerful, clear fluid. Great care must be taken with its use as a general anesthetic, however, because it has a narrow margin of safety. Even more powerful than chloroform and ether is a new synthetic general anesthetic, fluothane. It is injected into the body in measured amounts with a calibrated apparatus.

Two drugs often used as general anesthetics for operations of short duration are the liquids vinethene, which causes rapid anesthesia, and trilene,

[14]Nell Ann Pickett and Ann A. Laster, *Technical English*, 2nd. Ed. (New York: Harper & Row, Pub., 1975), pp. 101–2.

which produces a light, pain-killing effect. Trilene is usually combined with nitrous oxide and oxygen. Other important drugs used as general anesthetics are thiobarbiturates. They may be injected intravenously, but they are more commonly combined with inhaled anesthetics or with narcotic drugs. Thiopental sodium, or truth serum, is one example of a thiobarbiturate.

Not all surgery requires that the patient be unconscious. For minor operations, only a restricted, or local, area of the body need be made insensible to pain; thus a local anesthetic is administered. The local anesthetic prevents sensations of pain from traveling through the nerves in the drugged area.

Local anesthesia can be produced through three sites of injection. Infiltration is the injection of the drug into the tissues. Block anesthesia is produced by the injection of the drug around the main nerves leading to the operation area. These main nerves are blocked from transmitting sensory impulses. Spinal anesthesia results from the injection of the drug into the space surrounding the spinal cord. Because spinal anesthesia causes complete insensitivity to pain and muscular relaxation to the part of the body below the site of injection, it can be used only below the level of the cord that gives rise to spinal nerves aiding in breathing and heart action.

The most commonly used local anesthetics are the synthetic drugs Novocaine, Pontocaine, Metycaine, Nupercaine, and Xylocaine. These synthetic local anesthetics are far superior to the once widely used cocaine. Cocaine is no longer injected because it is too irritating to the tissues; its use is confined to topical application to mucous membranes.

The group of drugs designated anesthetics, in their power to cause complete or partial loss of feeling, is essential in modern surgery.

Section 9
ANALYSIS

The basic structure of a classification, discussed in Section 7, is as follows:

      I. Introduction
        A. Formal definition
        B. Purpose
        C. Value of classification (optional)
        D. Plan-of-development sentence
     II. Description of category A
    III. Description of category B
    IV. Description of category C
     V. Conclusion

This structure can be varied considerably. If the classification is simply a list of species, the plan-of-development sentence is often deleted, especially if the species are presented in outline form or with boldface subtitles. The conclusion is also deleted in some cases.

As described in Section 7, another way to write a classification is to present broad categories and then subcategories before describing the species. Such a format would look like this:

    I. Introduction
       A. Formal definition
       B. Plan-of-development sentence: broad categories A and B
       C. Description of broad categories
    II. Broad category A
       A. Division into subcategories 1 and 2
       B. Description of subcategories
   III. Detailed description of species in subcategory 1
   IV. Detailed description of species in subcategory 2
    V. Broad category B
       A. Division into subcategories 1 and 2
       B. Description of subcategories
   VI. Detailed description of species in subcategory 1
  VII. Detailed description of species in subcategory 2
 VIII. Conclusion

EXERCISE 9.1   With another student, preferably one who is in your field, analyze Model 1 in Section 8. Determine the function of each sentence in the model (formal definition, plan-of-development sentence, categories and sub-categories, etc.), and label each sentence in the margin. Discuss your results with your partner. Do you agree?

EXERCISE 9.2   Analyze Model 2 in the same way as in Exercise 9.1. Compare your analysis with that of Model 1. Are they different? If so, how?

Section 10
CHOOSING A TOPIC

Your assignment is to write a classification in your own field. First you must choose a topic. Many examples have been given in this unit that are good topics for a classification (e.g., food, scientific instruments, diseases). Here are some additional topics related to certain majors:

Biology/Plant Physiology	Civil/Mechanical Engineering
Cereal Grains	Foundations
Fungi	Machine Tools
Plant Diseases	Metals
Soil Types	Types of Dams

Electrical Engineering/ Computer Science	Chemistry/Chemical Engineering
Computer Languages	Alcohols
Computer Peripherals	Cooling Towers
Power Generation	Laboratory Glassware
Transistors	The Periodic Table of the Elements

Medicine/Physiology	Physics/Astronomy
Blood Vessels	Accelerators
Bones in the Skeleton	Forms of Electromagnetic Radiation
Carcinogens	Stars
Organs of the Body	Subatomic Particles
Vitamins	

EXERCISE 10.1   Write a classification in your own field using one of the structures shown in Section 9.

UNIT V
The Abstract

PART I
Grammar

Section 1
ARTICLES

The generic article was discussed in detail in Unit IV. Now we shall look at the effect of modification on the generic article.

THE ARTICLE IN PREMODIFIED GENERIC NOUN PHRASES

Premodified generic noun phrases are similar to premodified specific noun phrases: the addition of an adjective does not affect the choice of article.

Abstract generic:
The chip allowed the miniaturization of computers.
The silicon chip allowed the miniaturization of computers.
Concrete generic:
Chips are used in calculators and computers.
Silicon chips are used in calculators and computers.

THE ARTICLE IN POSTMODIFIED GENERIC NOUN PHRASES

Postmodified generic noun phrases almost always occur in their noun-compound forms in modern scientific English (see Unit VI, Section 3, "Noun Compounds"). These noun phrases (in both their abstract and concrete generic forms) can be divided into two classes: (1) descriptive and (2) partitive. The most common forms of postmodification are relative clauses, prepositional phrases, and *of*-phrases.

Descriptive Noun Phrases Postmodifed with Relative Clauses or Prepositional Phrases

Descriptive postmodified noun phrases simply describe or define the noun. They do not limit it to a certain quantity or amount. Look at the following examples.

Abstract descriptive. Abstract descriptive postmodified noun phrases occur only occasionally in scientific English:

> The drug that will cure cancer is the hope of both patients and researchers. (relative clause)
> The reactor out of control is symbolized by Three Mile Island. (prepositional phrase)

Concrete descriptive: countable

> A display unit that uses a cathode-ray tube may cause eyestrain. (relative clause, singular)
> Display units that use cathode-ray tubes may cause eyestrain. (relative clause, plural)
> A system for the purification of water removes dangerous microorganisms. (prepositional phrase, singular)
> Systems for the purification of water remove dangerous microorganisms. (prepositional phrase, plural)

Concrete descriptive: uncountable.

> Water that has been polluted is unfit to drink. (relative clause)
> Glucose in the urine indicates a variety of maladies. (prepositional phrase)

All of the above postmodified generic noun phrases can be transformed into noun compounds as follows:

> The anticancer drug is the hope of both patients and researchers.
> The uncontrollable reactor is symbolized by Three Mile Island.

A cathode-ray-tube display unit may cause eyestrain.
A water-purification system removes dangerous microorganisms.
Polluted water is unfit to drink.
Urine glucose indicates a variety of maladies.

Descriptive Noun Phrases Postmodified with *Of*-Phrases

Abstract descriptive. Abstract generic descriptive noun phrases postmodified with *of*-phrases are always in the form of noun compounds.

Incorrect	Correct
*the exchanger of heat	the heat exchanger
*the insulator of glass	the glass insulator
*the test of blood	the blood test

Concrete descriptive: singular and plural. Concrete generic descriptive noun phrases modified with *of*-phrases are quite common, although the noun compound can also be used.

An injection of morphine (a morphine injection) can relieve the pain of cancer. (singular)

Injections of morphine (morphine injections) can relieve the pain of cancer. (plural)

A layer of silicon (a silicon layer) is utilized in an integrated-circuit chip. (singular)

Layers of silicon (silicon layers) are utilized in an integrated-circuit chip. (plural)

Concrete descriptive: uncountable. Uncountable generic descriptive noun phrases postmodified with *of*-phrases usually sound old-fashioned. For example, *mercurous oxide* used to be called *oxide of mercury*. Here are some other examples:

Old term	Modern term
oil of vitriol	sulfuric acid
permanganate of potassium	potassium permanganate
oil of cinnamon	cinnamon oil
slake of lime	calcium hydroxide

However, some of these old-fashioned terms are still used in reference to specific substances:

plaster of Paris (a white paste made of gypsum and used for casts)
mother of pearl (the substance lining the shell of oysters, etc.)
essence of musk (a perfume containing a substance secreted by the male musk deer)

EXERCISE 1.1   The underlined phrases in the following sentences are descriptive generic phrases. Add *the* to the blank if the phrase is abstract generic. Add *a(n)* or *0* if the phrase is concrete generic. Sometimes more than one answer is possible.

1. _____ amalgam of 50 percent mercury and 50 percent silver is used in dental fillings.
2. _____ fuel that is inexhaustible, many scientists believe, is hydrogen.
3. _____ essence of roses is used as perfume.
4. _____ systematic name of an organic compound shows its chemical structure.
5. _____ nitrogen that has been irradiated glows.
6. _____ experiments in wind tunnels study the reaction of models to stress.
7. _____ air in a water pipeline can cause a knocking sound.
8. _____ white dog in front of the old-fashioned phonograph is a symbol for a certain electronics company.
9. A quasar is _____ starlike object that emits light and microwave radiation.
10. _____ birds of prey help to control the rodent population.

Partitive Postmodified Generic Noun Phrases

Partitive postmodified generic noun phrases restrict the quantity or amount of the noun with an *of*-phrase.

Abstract partitive.   Abstract partitive postmodified noun phrases are rare. Only postmodification with an *of*-phrase is possible, and this is more likely to occur in literary than in scientific English.

The cup of tea is a symbol of British gentility.

Concrete partitive: singular and plural.   Concrete partitive noun phrases postmodified with *of*-phrases are common:

A pack of cigarettes a day can cause lung cancer. (*of*-phrase, singular)
Two packs of cigarettes a day can cause lung cancer. (*of*-phrase, plural)

Concrete partitive: uncountable.   Uncountable partitive concrete noun phrases are not possible because the head noun in a partitive phrase must be a distinct shape, a measured quantity, or a container (all countable nouns):

a grain of sand
a kilogram of sand
a bucket of sand

EXERCISE 1.2   Underline the partitive generic noun phrases in the following sentences. Add *a(n)* or *∅* to the blanks.

1. _____ bolt of lightning is often used as the symbol for electrical danger.
2. _____ tiny bubbles of air damage the metal in water turbine blades.
3. Erupting volcanoes sometimes produce_____ rivers of lava.
4. The human body has_____ standard temperature of 98.6° F (37.0° C).
5. _____ small quantity of uranium is required for an atomic bomb.
6. _____ stream of ions is produced by a cyclotron and used to help cancer patients.
7. _____ mole of any element contains 6.02×10^{23} atoms.
8. _____ reservoirs of water are known to exist under the Sahara Desert.
9. _____ cylinder of compressed propane supplies gas in remote places.
10. _____ block of ice has a slightly larger volume than an equivalent mass of water.

Limited Generic Reference: *the*

In Unit III, Section 1, we saw that the article *the* limits a specific uncountable or plural noun only when it is postmodified. Some examples are:

The wood that the carpenter bought is dry now.
The trees that the gardeners planted have all produced fruit.

This limitation with *the* also applies to concrete generic uncountable and plural countable nouns. It is called limited generic reference, and it is the only concrete generic form that uses the article *the* (with the exception of second-mention generic occurrences). Look at these examples:

1. The water that comes from rivers sometimes carries typhoid. (relative clause, uncountable)
2. The cars that come from Japan are very well made. (relative clause, plural countable)
3. The soil from deciduous woods is good for crops. (prepositional phrase, uncountable)
4. The animals in most underground caves are blind and colorless. (prepositional phrase, plural countable)
5. The wine of France is a product of art and science. (*of*-phrase, uncountable)
6. The wines of France are products of art and science. (*of*-phrase, plural countable)

Notice that the postmodifying phrase **limits** the noun, usually by indicating the source or location of that noun. In sentence 1, we mean not **any water** but only the water that comes from rivers. These phrases are partitive in the sense that they indicate a part of the whole (a glass of water, a river of water). If we remove the postmodifying phrase, we cannot use *the:*

> Water sometimes carries typhoid.
> Soil is good for crops.
> Wine is a product of art and science.

If the postmodifying phrase indicates an alteration or a process that the noun it modifies has undergone, then the phrase is descriptive rather than partitive and is not an example of limited generic reference. It is incorrect to use *the* before such a phrase (unless the phrase is a second-mention occurrence).

> Incorrect:
> *The DNA that has been exposed to ultraviolet radiation is sometimes unable to replicate.
> *The animals that become infected with rabies must be killed.

> Correct:
> DNA that has been exposed to ultraviolet radiation is sometimes unable to replicate.
> Animals that become infected with rabies must be killed.

Keep in mind that many native speakers do not use *the* in sentences like 1–4 on page 201, even though it is always correct to do so:

> Water that comes from rivers sometimes carries typhoid.
> Cars that come from Japan are very well made.
> Soil from deciduous woods is good for crops.
> Animals in underground caves are blind and colorless.

However, this deletion is never possible with *of*-phrase postmodification, as in sentences 5 and 6 on page 201.

> Incorrect:
> *Wine of France is a product of art and science.
> *Wines of France are a product of art and science.

EXERCISE 1.3   The underlined phrases in the following sentences are first-mention generic nouns postmodified with relative clauses or prepositional phrases. (1) If the postmodifier indicates a source or location, suggesting that there are different types of the noun it modifies, label the sentence partitive (P). (2) If the postmodifier indicates an alteration or a process that the noun it modifies has undergone but does not suggest different types of the noun it modifies, label the sentence descriptive (D). (3) Insert the correct article before the modified noun. In some cases, more than one answer is possible.

Example: _____ _____ oil that comes from Saudi Arabia is of a very high quality.
Answer: __P__ The (∅) oil that comes from Saudi Arabia is of a very high quality.

1. _____ _____ oil that comes from the kikui nut is used for curing wood.
2. _____ _____ nitrogen in the soil is replaced by special bacteria that live on roots.
3. _____ _____ books whose pages are bent are called *dog-eared*.
4. _____ _____ oil that has been used for frying is highly saturated.
5. _____ _____ diseases caused by viruses are difficult to cure.
6. _____ _____ wood that comes from the balsa tree is used for making models.
7. _____ _____ tanks that store petroleum must be periodically cleaned.
8. _____ _____ books that were published in the sixteenth century are difficult to read.
9. _____ _____ wood that is produced in dry areas is often brittle.
10. _____ _____ water containing raw sewage is dangerous to wildlife.
11. _____ _____ wood that has not been allowed to dry sufficiently cannot be painted or varnished.
12. _____ _____ blood that has not been properly screened should not be used for transfusions.
13. _____ _____ oil that is pressed from peanuts is good for cooking.
14. _____ _____ blood from an insect is different from human blood.
15. _____ _____ cattle that have been immersed in a dipping vat are protected from ticks and other parasites.

EXERCISE 1.4   Add *a(n)*, *the*, or *∅* to the following passage where necessary.

ANEROID BAROMETER

Aneroid barometer is instrument that depends on changing volume of container to indicate atmospheric pressure. It consists of airtight box of thin flexible material with air inside partially evacuated. One side of evacuated box is attached to spring. When atmospheric pressure increases, box tends to collapse. When atmospheric pressure decreases, sides of box spring outward. This slight movement is magnified by series of levers connected to indicator needle that shows atmospheric pressure.

Titles and labels, like many newspaper and magazine headlines, often use "telegraphic speech." Telegraphic speech, the limited language people use when sending a telegram (because they must pay for each word), regularly deletes any words that are not absolutely necessary for understanding (i.e., most of the function words, including the articles, prepositions, and conjunctions). In titles and labels, however, only the articles are affected. Look at these examples:

> Figure title: Solubility of carbon disulfide with methyl alcohol
> Longitudinal section of hair follicle
> Map title: Average rainfall and temperature (in degrees F) of South America
> Photo label: Shedding of skin by red-bellied water snake

If we replace the deleted articles, these titles and labels become complete noun phrases.

> <u>The</u> solubility of carbon disulfide with methyl alcohol
> <u>A</u> longitudinal section of <u>a</u> hair follicle
> <u>The</u> average annual rainfall and temperature (in degrees F) of South America
> <u>The</u> shedding of <u>the</u> skin by <u>the</u> red-bellied water snake

Unfortunately, there are no precise rules for article deletion in titles and labels. It is usually a matter of space. In most cases the first article is deleted. In many cases, all articles are deleted (see examples above), but there are also many cases where a subsequent article is retained:

> Large valley on Tethys may be <u>a</u> crack caused by impact on opposite side (photo label)
> Right pectoral girdle and fin of <u>an</u> extinct crossopterygian fish (figure title)
> Range in <u>the</u> Andes Mountains, central Chile (photo label)[1]

EXERCISE 1.5   Add the articles *a(n)* or *the* to the following titles and labels.

Example: hexagon-head bolt (figure title)
 Answer: a hexagon-head bolt (figure title)

> 1. Head of pit viper, showing poison and sense mechanisms (figure title)
> 2. Formation of radical is rate-controlling step (figure title)

[1]Reprinted with permission from *Science News,* the weekly newsmagazine of science, copyright 1983 by Science Service, Inc.

3. Effect of temperature on growth of psychrophillic *Bacillus* sp. (figure title)
4. Smoking is hazardous to heart (title)
5. Electrode picks up signal from individual fiber, 1 to 15 microns in diameter, within nerve (photo label)
6. Beam bolted at both ends with distributed load (figure title)
7. Photomicrograph of transistor from integrated circuit shown in Fig. 7.2 (photo label)
8. Thermal barrier: design of stainless-steel thermal barrier between pump and motor permits operation regardless of time between deliveries (catalog description)
9. Synthesis of specific antibody following exposure to antigen (title)
10. Address block on magnetized drum showing how photoelectric cells magnetize spots on drum to match holes on cards or tape (photo label)

EXERCISE 1.6　How would you delete the articles in the following titles and labels if you were going to use them in a report?

Example:　A high-speed printer (photo label)
Answer:　High-speed printer (photo label)

1. A turbine-driven auxiliary oil pump (figure title)
2. A programmer using a template (photo label)
3. The hazards of multicolor laser beams (title)
4. A patient connected to a kidney-dialysis machine (photo label)
5. The flow of ions through a single channel in a membrane can be detected (title)
6. The ALCO 510 turbosupercharger rotor assembly: an exploded view (figure title)
7. The printer can be stopped for the insertion of text or forms alignment (catalog description)
8. Plant layouts are designed for maximum efficiency by adapting the flow of manufacturing operations to the available space (photo label)
9. Tree rings keep a record of the climatic conditions through the centuries (photo label)
10. A schematic sketch of the flow around a bend in a rectangular channel (figure title)

Section 2
SENTENCE COMBINING: PARTICIPIAL PHRASES

Participial phrases are *-ing* (active) and *-ed* (passive) phrases that are based on verbs. They can have either adjectival or adverbial functions. We studied participial phrases in Units II and III, Section 3, when we discussed the reduction of the relative pronoun + *be* in subject-form relative clauses.

Active: The weather system (that is) <u>approaching the coast</u> is a hurricane.
Passive: The plutonium (that was) <u>stolen from the lab</u> was never found.

In Unit III, Section 4, we discussed the infinitive form without *to* after perception verbs. The continuous form of such a structure is also an example of an adjectival participial phrase.

The engineer saw the oil <u>floating near the platform</u>.

Sentences with participial phrases are produced by sentence combining and, like the TIME group of coordinators and subordinators, indicate either simultaneous or sequential actions.

PARTICIPIAL PHRASES SHOWING SIMULTANEOUS ACTIONS

Simultaneous actions are those that happen at the same general time. Look at the following example:

Fleming was studying the deadly staphylococcus.
+ Fleming discovered penicillin.

(1) When Fleming was studying the deadly staphylococcus, he discovered penicillin.
(2) Because Fleming was studying the deadly staphylococcus, he discovered penicillin.

Sentence 1 is the result of sentence combining using the TIME subordinator *when* (we could also have used *while* or *as*). Sentence 2 is the result of sentence combining using the SO subordinator *because* (we could also have used *since* or *as*). We studied these in Unit IV, Section 2.

When the subject of the two sentences is the same (in this case, *Fleming*), we can also combine the sentences using a simultaneous participial phrase.

(3) <u>Studying the deadly staphylococcus</u>, Fleming discovered penicillin.

Time and cause-effect relationships sometimes overlap, the time sequence often implying cause and effect. The participial phrase retains the obscurity of the relationship; in other words, we often cannot tell whether a participial phrase indicates a cause or simply a time. From sentence 3, we do not know if

Fleming discovered penicillin **because** he was studying the staphylococcus or simply **while** he was doing so.

Simultaneous participial phrases can also be passive:

The laboratory is funded by the federal government.
+ The laboratory designs nuclear weapons.

Funded by the federal government, the laboratory designs nuclear weapons.

Notice that the subject *(laboratory)* must be the same in each sentence if we use a participial phrase.

EXERCISE 2.1   Combine the following pairs of sentences by making sentence (a) into an active or passive simultaneous participial phrase.

Example: (a) The Suez Canal was completed in 1869.
         (b) The Suez Canal is 100.6 miles long.
Answer: Completed in 1869, the Suez Canal is 100.6 miles long.

   1. (a) The fire burned at a temperature of 1800° F.
       (b) The fire destroyed most of the building.
   2. (a) Australia was separated from Asia long ago.
       (b) Australia has many unusual plants and animals.
   3. (a) Venus is covered by thick clouds.
       (b) Venus has a surface temperature of 800° F.
   4. (a) The heart beats at an average of seventy times a minute.
       (b) The heart keeps blood moving continuously during life.
   5. (a) The plane flew at the speed of sound.
       (b) The plane produced a shock wave called a sonic boom.
   6. (a) The throttle valve is opened and closed by the accelerator pedal.
       (b) The throttle valve controls the amount of fuel that enters the cylinder.
   7. (a) Uranium waste decays to lead in 4.5 billion years.
       (b) Uranium waste presents a serious disposal problem.
   8. (a) FORTRAN is based on mathematical principles.
       (b) FORTRAN is used primarily for solving scientific problems.
   9. (a) Gravitational force depends on the mass of the bodies involved.
       (b) Gravitational force exists throughout the universe.
  10. (a) The orchid is grown in a temperate climate.
       (b) The orchid is the most highly organized flower among the cotyledons.

EXERCISE 2.2   Transform the following cause-and-effect sentences into sentences with participial phrases.

Example: The plant did not grow because it had too little light.
Answer: Having too little light, the plant did not grow.

1. The building collapsed because it had a weak foundation.
2. Alkanes consist entirely of single bonds; therefore, they are very stable and resistant to chemical change.
3. The animals needed water, so they migrated to the mountains.
4. Carbon is the basis of all organic compounds when it is combined with hydrogen and oxygen.
5. The farmer wanted to kill the weeds in his cornfield; for this reason, he plowed the field and used a selective weed killer.
6. Milk sours easily; therefore, it should be kept in the refrigerator.
7. The tomatoes grew rapidly because they were grown in a hydroponic solution.
8. Because it was built with insufficient wind bracing, the original Tacoma Narrows Bridge was very unstable.
9. If a piece of graphite is subjected to tremendous heat and pressure, it becomes a diamond.
10. Ionized atoms of silver are deposited on the electroplated object and replaced by others dissolved from a silver bar at the positive terminal.

PARTICIPIAL PHRASES SHOWING SEQUENTIAL ACTIONS

Sequential actions are those that come before or after each other. Look at the following example:

The geochemist has extracted the lead 204 from the rock sample.
+ The geochemist can determine the rock's age.

_____

(1) After the geochemist has extracted the lead 204 from the rock sample, he can determine its age.
(2) Because the geochemist has extracted the lead 204 from the rock sample, he can determine its age.

Sentence 1 is the result of sentence combining with the TIME subordinator *after*. Sentence 2 is the result of sentence combining with the SO subordinator *because*. Since the subject of the two sentences *(the geochemist)* is the same, we can also combine the two sentences using a sequential participial phrase:

(3) <u>Having extracted the lead 204 from the rock sample</u>, the geochemist can determine its age.

Sequential participial phrases can also be passive:

> Electrical power is produced at the dam.
> + Electrical power is transported by high-voltage lines.

Having been produced at the dam, electrical power is transported by high-voltage lines.

Notice that the sequential participial phrase has a different structure from the simultaneous participial phrase:

VERB: EXTRACT

FORM	ACTIVE	PASSIVE
Simultaneous	extracting	extracted
Sequential	having extracted	having been extracted

EXERCISE 2.3   Combine the following pairs of sentences by making sentence (a) into an active or passive sequential participial phrase **if possible**.

Example: (a) We determine the radius and height of a cylinder.
(b) We can calculate the volume.
Answer: Having determined the radius and height of a cylinder, we can calculate the volume.

1. (a) Geologists sank a pipe 1000 feet down into the glacier.
   (b) Geologists were able to measure tne difference between surface and deep flow.
2. (a) Sewage is filtered.
   (b) Sewage is subjected to biosynthesis for further purification.
3. (a) The gas expands suddenly.
   (b) The piston is forced into the cylinder.
4. (a) The hydrogen within a star reaches a temperature of 20 million degrees F.
   (b) The hydrogen within a star ignites and burns in a continuing series of nuclear reactions.
5. (a) The soil is treated with a disease-controlling chemical.
   (b) The soil is allowed to air for a few days before use.
6. (a) The slide was prepared in a water bath.
   (b) The slide was dried in an incubator at 37° C.
7. (a) A mineral-dust particle meets supercooled vapor in extreme cold.
   (b) A mineral-dust particle becomes a snow crystal.
8. (a) The dinosaurs ruled the earth for 100 million years.
   (b) The dinosaurs became extinct, opening the way for the mammals.

9. (a) The compiler translates a program into machine language.
   (b) Millions of switches are activated in a computer during processing.
10. (a) A zinc machine part is formed by cold die-casting.
    (b) A zinc machine part must be finished to remove casting seams.

V_{ING} CLAUSES OF RESULT

We can use different grammatical structures to describe a series of events. A common sentence structure in scientific writing uses (1) a subordinate clause, (2) a main clause, and (3) a simultaneous participial phrase to express result. Look at this example:

(1) A plane reaches the speed of sound.
(2) Air molecules are compacted into a dense layer.
+(3) This condition causes a sonic boom.

When a plane reaches the speed of sound, air molecules are compacted into a dense layer, causing a sonic boom.

Notice that the subject of sentence 3 is the entire sentence 2, not just the subject, as in other participial phrases. The fact that the participial phrase is a result can be emphasized by placing a "result" word before it (although this is **not** required):

Air molecules at the nose of a plane flying at the speed of sound are compacted, <u>thus</u> causing a sonic boom.

"Result" words similar to *thus* include *thereby, so,* and *in this manner.* Examples:

The meteorite hit the earth, <u>thereby forming a large crater</u>.
The cell divides, <u>so producing a blastomere</u>.
Solar-wind particles strike molecules in the atmosphere, <u>in this manner creating the aurora borealis</u>.
The galaxies collided, <u>generating 10^{36} watts of radio power</u>.

EXERCISE 2.4   Combine the following sentences, making the second sentence into a V_{ing} clause of result.

Example: The earthquake struck. This caused several buildings to collapse.
 Answer: The earthquake struck, causing several buildings to collapse.

1. The coal-gasification plant had financial troubles. This left the project's future uncertain.
2. Blood filtering has been found to have no effect on schizophrenia. This contradicts an earlier report.

3. Scientists have found new methods of breaking the bonds of hydrocarbons such as petroleum. This makes it possible to manipulate the chemicals into drugs and other complex materials.
4. Certain insect viruses are genetically altered. This produces large amounts of human interferon.
5. The liquid outer portion of the earth's core (about 95 percent) is constantly in motion. This causes the earth to have a magnetic field.
6. There have been reports of unauthorized entries into computer systems. This raises concerns about computer security.
7. Cigarette smoking exposes tissue to high concentrations of formaldehyde. This possibly leads to respiratory-tract cancers.
8. The sun will probably swell to a red giant in 5 billion years. This will vaporize the earth and any creatures that are left on its surface.
9. The identities of neutrons and protons in certain atomic nuclei may break down. This turns the nuclei into mixtures or plasmas of quarks and gluons.
10. The blood of severely alcoholic men contains a substance not found in the blood of social drinkers. This indicates that alcoholics may use a unique physiological pathway for the breakdown of ethanol.

DANGLING MODIFIERS

A common mistake in using participial phrases is to forget that the subject of the participial phrase must be the same as the subject of the main clause. If the subjects are different, the result is a dangling modifier, an incorrect grammatical structure. Look at this example:

(a) *Being over 2 million years old, the professor placed the fossil in the box with great care.
(b) *Preserved in formaldehyde, Dr. White washed the frog before dissecting it.

The first sentence is incorrect because it implies that the professor is over 2 million years old. The second is incorrect because it implies that Dr. White was preserved in formaldehyde. The easiest way to correct these sentences is to change the position of the subject in the main clause.

(1a) Being over 2 million years old, the fossil was placed in the box (by the professor) with great care.
(1b) Preserved in formaldehyde, the frog was washed before Dr. White dissected it.

In some cases, we can correct the sentence by simply adding a subject to the participial phrase. If a subject is present in the participial phrase, the subject of the main clause does not have to be the same.

> (2a) <u>The fossil</u> being over 2 million years old, <u>the professor</u> placed it in the box with great care.
> (2b) <u>The frog</u> preserved in formaldehyde, <u>Dr. White</u> washed it before he dissected it.

The use of *being* in sentence 1a is correct only because it implies cause and effect. In other words, we could state (1a) as follows:

> (3a) The fossil was placed in the box (by the professor) with great care **because** it was 2,000,000 years old.

Being should not be used if there is no cause and effect:

> Incorrect:
> *Being over 2 million years old, the fossil was found in Egypt.
> Correct:
> Over 2 million years old, the fossil was found in Egypt.

In very formal writing, *being* can be deleted from cause-and-effect sentences:

> (4a) Over 2 million years old, the fossil was placed in the box with great care.

Here are some other examples of dangling modifiers. It is a good idea to carefully check any sentence beginning with a V_{ing} or a V_{ed} form to see if it is a dangling modifier.

> *Before going to bed, cigarettes should be extinguished.
> Before going to bed, smokers should extinguish their cigarettes.
> *Working day and night, the dam was completed in less than two years.
> Working day and night, the laborers completed the dam in less than two years.
> <u>Two thousand people</u> working day and night, the dam was completed in less than two years.

Notice that the last example has a subject as well as a V_{ing} form. If such a subject is present, the subject of the V_{ing} form does **not** have to be the same as the subject of the main clause:

> *Located near the stomach, the patient said the tumor was making her feel nauseated.

Located near the stomach, the tumor was making the patient feel
  nauseated.
<u>The tumor</u> being located near the stomach, the patient said she felt
  nauseated.

Sometimes the best solution is to use an adverbial phrase instead of a participial phrase:

<u>As the fossil was over 2 million years old</u>, the professor carefully placed
  it in a box.
<u>Since the frog was preserved in formaldehyde</u>, Dr. White washed it be-
  fore he dissected it.
<u>Because the work crew worked day and night</u>, the dam was finished in
  less than two years.
<u>Before smokers go to bed</u>, they should extinguish their cigarettes.
<u>Since the tumor was located near the stomach</u>, it was making the pa-
  tient feel nauseated.

EXERCISE 2.5   Correct the following dangling-modifier errors by changing
the main sentence **if necessary**.

Example: Pushed up by forces in the earth's crust, water then erodes mountains into
  valleys and gorges.
  Answer: Pushed up by forces in the earth's crust, mountains are then eroded into
  valleys and gorges.

1. Being a planet with virtually no oxygen, there is little chance of
   life on Venus.
2. Injected into the combustion chamber of the gas turbine engine,
   the fuel ignites, and the resulting blast of hot gases drives the
   turbine.
3. Used for irrigation projects in the southwestern U.S., Mexico de-
   rives very little water from the Colorado River.
4. Their tops leveled by high, relatively stable air layers, thunder-
   heads forming from summer cumulus clouds take the shape of
   giant anvils.
5. Having developed a flow chart, a computer program is easier to
   construct.
6. Having sown rice grain by hand in a muddy field and let it germinate, a rice farmer transplants the young rice plants to a
   plowed, submerged area called a paddy field.
7. Having been formed ages ago when magma cooled, granite commonly occurs in mountain ranges or level regions that were moun-
   tainous at one time.
8. Contained in the abundant mineral known as gypsum, cold water
   dissolves calcium sulphate better than hot water.

9. Belonging to the same section of the animal kingdom as jellyfish and sea anemones, corals were long mistaken for plants because of their blossomlike appearance.

10. Differing from the adult stage, the eyes of insect larvae are simple, not compound.

EXERCISE 2.6   Correct the incorrect sentences in Exercise 2.5 by changing the participial phrase rather than the main sentence. There may be more than one way to do this.

Example: Pushed up by forces in the earth's crust, rain then erodes mountains into valleys and gorges.

Answers: After mountains are pushed up by forces in the earth's crust, rain erodes them into valleys and gorges.

Forces in the earth's crust having pushed up mountains, rain then erodes them into valleys and gorges.

V$_{ING}$ CLAUSES AS NOUN PHRASES

An entire sentence can be transformed into a noun phrase by changing the verb to the V$_{ing}$ form and making the subject possessive. Look at this example:

(a) X = A plane   reaches   the speed of sound.

(A plane's) (reaching)

+ (b) X causes a sonic boom.

(1) A plane's reaching the speed of sound causes a sonic boom.

Notice that we transform the subject of sentence (a) into the possessive form and the verb into the V$_{ing}$ form. Sentence (a) thus becomes the complete subject of sentence 1. The bare subject—that is, the noun that controls the number of the verb—is the V$_{ing}$ form, which is always singular. Here are some additional examples:

The meteorite's hitting the earth formed a large crater.
The galaxies' colliding generated 10^{36} watts of radio power.
The cell's dividing has produced a blastomere.
Solar-wind particles' striking molecules in the atmosphere creates the aurora borealis.

The possessive V$_{ing}$ form occurs most commonly as the subject of a sentence. Possessive V$_{ing}$ forms can also occur as objects. Look at these examples:

V$_{ing}$ form as direct object:
Many hospitals discourage women's having their babies at home.

V_{ing} form as object of preposition:
A sonic boom is caused by <u>a plane's reaching the speed of sound</u>.

We must be careful not to confuse a possessive V_{ing} form with a reduced subject-form defining relative clause (see Unit II, Section 3, "Reducing Defining Relative Clauses"). Look at these examples:

Possessive V_{ing} form:
SUBJECT
<u>A plane's reaching the speed of sound</u> causes a sonic boom.

Reduced subject-form defining relative clause:
SUBJECT
A plane <u>reaching the speed of sound</u> causes a sonic boom.

The possessive V_{ing} form indicates that the event of reaching the speed of sound caused the sonic boom. The relative clause indicates that the plane itself caused the sonic boom.

EXERCISE 2.7   Combine the following pairs of sentences, making sentence (a) into a possessive noun phrase.

Example: (a) The electrons spin around the nucleus.
(b) This determines the "shape" of an atom.
Answer: The electrons' spinning around the nucleus determines the "shape" of the atom.

1. (a) An object falls into a black hole.
   (b) This causes strong X rays to be released.
2. (a) The operation will be rescheduled for Wednesday.
   (b) This will improve the patient's chances of recovery.
3. (a) The moon orbits around the earth.
   (b) Tides are partially produced by this.
4. (a) A mammal takes care of its young.
   (b) This is an example of evolutionary progress.
5. (a) The volcano erupted explosively.
   (b) This caused volcanic ash to be deposited for miles around.
6. (a) A pelican flies low over the water.
   (b) This allows it to see fish for prey.
7. (a) Insects carry pollen from flower to flower.
   (b) The reproduction of many flowering plants depends on this.
8. (a) An object moves at the speed of light.
   (b) This causes time to stop relative to that object.
9. (a) Water flows at high velocity around propellers and turbines.
   (b) This causes the severe damage known as cavitation.
10. (a) Cholesterol builds up on the walls of arteries.
    (b) Heart attacks can be triggered by this.

Section 3
MODIFICATION AND REDUCTION

ADJECTIVE COMPOUNDS

An adjective compound is an adjective that consists of two or more words connected by a hyphen. Look at this example:

(1) The geologist collected a <u>rare-earth</u> sample.

This sentence indicates that the geologist collected a representative of one of the elements known as *rare earths* (elements with numbers 57–71, also known as the lanthanide series). It is also possible to write sentence (1) without a hyphen.

(2) The geologist collected a <u>rare earth</u> sample.

Sentence (2) has a different meaning. It states that the geologist collected a sample of earth that was rare (not commonly found). Thus we see that the hyphen in an adjective compound indicates a significant difference in meaning. Normally, we interpret the last two words *(earth sample)* as a noun compound (see Unit VI, Section 3). The hyphen prevents us from interpreting the last two words as a noun compound. Here are some other examples:

a small car factory = a car factory that is small
a small-car factory = a factory that makes small cars
the red star spectrogram = the star spectrogram that happens to be red
the red-star spectrogram = the spectrogram of a red star
productive time estimates = time estimates that are productive or useful
productive-time estimates = estimates of productive time as compared
    with nonproductive time

The tendency in modern scientific writing is to use the hyphen for adjective compounds only. However, older texts, especially British ones, often use hyphens between the nouns in a noun compound as well.

Adjective compounds without a corresponding nonhyphenated form are more common. They express several relationships, as shown in the following examples:

1. measurement
   a three-meter cable (the cable is three meters long)
   a 200-horsepower motor (the motor produces 200 horsepower)
   a 110-volt outlet (the outlet delivers 110 volts)
2. two substances
   an oil-water suspension (the suspension contains oil and water)
   an O_2-NO_2 mixture (the mixture consists of oxygen and nitrous oxide)

a copper-constantan thermocouple (the thermocouple is made of copper and constantan)

3. attributes
    a U-shaped tube (the tube has a U shape)
    a five-sided polygon (the polygon has five sides)
    a flat-headed screw (the screw has a flat head)

4. a verb-object relationship
    sulfur-containing additives (the additives contain sulfur)
    all-inclusive rules (the rules include all or everything)
    a self-regulating system (the system regulates itself)
    CO_2-generating bacteria (the bacteria generate CO_2)
    a time-consuming method (the method consumes time)
    a cholesterol-blocked artery (the artery is blocked by cholesterol)

5. a verb-adverb relationship [*Note:* some writers do not consider the hyphen to be necessary in this construction.]
    a widely-used substance (the substance is used widely)
    a fully-developed program (the program is developed fully)
    a little-known region (the region is not known well)

6. combinations
    a fifty-cycle, high-pass filter
    a thin-film, metal-base transistor
    a word-identity and card-location code

EXERCISE 3.1 Make the following phrases into adjective compounds.

Example: a pump is driven by gears
 Answer: a gear-driven pump

1. the highway is 5000 kilometers long
2. a reaction is induced by drugs
3. the complex contains copper and ammonia
4. an artery has thick walls
5. this method is used widely
6. a figure has three sides
7. the sandstone bears oil
8. the truck weighs five tons
9. chemicals are constructed in a synthetic manner
10. the city air is free of smoke and low in particulates

EXERCISE 3.2 Add a hyphen to the following sentences where necessary.

Example: The machine is based on a gear driven pump.
 Answer: The machine is based on a gear-driven pump.

1. Einstein described the space time continuum.
2. Sutter Labs is an industrial chemistry laboratory.
3. A one kilowatt power amplifier has been installed.

4. A trimaran is a three hulled boat.
5. Cholorofluorocarbons disrupt the ozone layer chemistry.
6. Dioxin is thought to be a cancer causing substance.
7. Use only a glass microscope slide.
8. The pulley has a V shaped notch.
9. The device utilizes a constant level speech processing system.
10. This circuit requires thin film, metal base transistors.

PREPOSITIONS OF PLACE

The prepositions of place describe relative position (stationary) and passage or direction (moving).

Relative Position

In Unit III, we discussed the uses of *at* to indicate a position or a location that implies a function, *on* to indicate a surface or a line, and *in* to indicate containment or mode. We will now look at place prepositions that indicate (1) a horizontal relation, (2) a vertical relation, and (3) a volumetric relation.

Horizontal.   Horizontal directions are described by prepositions indicating position and proximity.

1. position
   a. line: *behind, between, in front of, along*

A -----B ----→C   Examples

A is behind B.   The oxygen tank is behind the
                                    building.
B is between A and C.   The earth's mantle is between the
                                    crust and the core.
C is in front of B.   Hold the chart in front of the
                                    patient.
A, B, and C are along the   There were cracks all along the fault
same line.   line.

   b. group: *among, amid, surrounded by, around*

B
A   E   C   Examples
D

E is among A, B, C, and D.   Monkeys live among the trees.
E is amid A, B, C, and D   Maggots live amid rotten flesh.
E is surrounded by A, B,   An island is surrounded by water.
C, and D.

| A, B,C, and D are around E. | The petals are around the center of the flower. |

2. proximity: *against, next to, beside, by, close to, near, not far from, in the neighborhood of, far from, a great distance from, nowhere near*

Examples

| Ⓐ Ⓑ A is against B. | The ladder is against the wall. |

| Ⓐ   Ⓑ A is next to B.
 A is beside B.
 A is by B. | The heart is next to the lungs.
 The xylem is beside the phloem.
 Crabs live by the ocean. |

| Ⓐ   Ⓑ A is close to B.
 A is near B. | The eyes are close to the brain.
 The switch is near the door. |

| Ⓐ   Ⓑ A is not far from B. | Coral lives not far from the surface. |

A is in the neighborhood of B.
Ⓐ

Fossils were found in the neighborhood of the polar ice cap.
                           Ⓑ

A is far from B.
A is a great distance from B.
A is nowhere near B.

Pluto is far from the sun.
The earth is a great distance from the galaxy's center.
The palms are nowhere near the snow line.

EXERCISE 3.3   Add the appropriate preposition of horizontal relative position to the blanks. There is one sentence for each preposition in the list below.

Example: The ladder is _____ the wall.
 Answer: The ladder is against the wall.

a great distance from	around	close to	next to
against	behind	far from	not far from
along	beside	in front of	nowhere near
amid	between	in the neighborhood of	surrounded by
among	by	near	

1. The nucleus of a cell is _____ the cytoplasm.
2. Magnesium is _____ sodium in the periodic table.
3. The retina of the eye is _____ the lens.
4. In astronomical terms, the sun is _____ the earth.
5. A new tunnel was constructed _____ the old one because of increased traffic.
6. Precipitation formed _____ the cold water pipes.

7. During a solar eclipse, the moon is actually _____ the earth.
8. The center of the Milky Way galaxy is _____ its outer edge.
9. The test was inaccurate: the tumor was _____ the predicted location.
10. The asteroid belt lies _____ Mars and Jupiter.
11. Many species of birds and fish travel _____ their place of birth and then return to bear their young.
12. The researcher lived _____ the chimpanzees for over fifteen years.
13. The liver is situated _____ the stomach.
14. Rust had developed all _____ the outer edge of the I-beam.
15. The tumor was pressing _____ the patient's stomach.
16. If recorded tape is placed _____ a magnet, the recorded material may be destroyed.
17. Willow trees usually live _____ rivers or streams.
18. Many microorganisms thrive _____ the mud and decayed matter at the bottom of a pond.
19. A volcano is usually found _____ a subduction zone, the area where one continental plate slides under another.

Vertical. Vertical directions are described by prepositions indicating contiguity and noncontiguity.

1. contiguity (touching): *on, on top of, underneath, beneath, all over*

Examples

[A]	A is on B.	The beaker is on the table.
[B]	A is on top of B	Dr. Dinn's report is on top of the pile.
[C]	C is underneath B.	Muscle is underneath the skin.
	C is beneath B.	The mantle is beneath the crust.
[B][A]	A is all over B.	The patient had scars all over his back.

2. noncontiguity (separated): *above, over, off, below, under*

Examples

[A]	A is above B.	Ozone forms above the clouds.
	A is over B.	Center the drill over the hole.
[B]	B is off D.	Oil was found off the shore.
[D]	D is below B.	Algae lives below the surface.
[C]	C is under B.	Keep the plant under the light.

Volumetric. Volumetric relations are indicated by three prepositions: *inside,* *outside,* and *throughout.*

Examples

A B	A is inside B.	The virus lives inside the cell.
A | B	A is outside B.	The male lion stays outside the den.
A B A A B	A is throughout B.	Cancer often spreads throughout the tissue it infects.

EXERCISE 3.4   Add the appropriate preposition of vertical or volumetric relative position to the blanks. There is one sentence for each preposition in the list below.

Example: The flask is _____ the table.
Answer: The flask is on the table.

above	inside	on top of	throughout
all over	off	outside	under
below	on	over	underneath
beneath			

1. Many plants will not grow _____ eucalyptus or redwood trees.
2. The oxide was removed to reveal the shiny metal _____.
3. The oil rig stands firmly _____ the ocean floor.
4. Giant sponges had grown _____ the leaking nuclear-waste container, completely covering it.
5. No trees live _____ the snow line.
6. The premature baby had to be placed _____ an incubator.
7. Subcutaneous fat is located _____ the skin.
8. Scientists describe what lies _____ the mantle of the earth by using artificial shock waves.
9. Hydrogen appears to exist _____ the universe.
10. Volatile materials should never be placed _____ heaters or furnaces.
11. A centrifuge becomes unstable if it is _____ center.
12. Most harmful bacteria cannot survive _____ a living body.
13. The mushroom cloud of an atomic bomb forms _____ the point of detonation.

Passage or Direction

Prepositions of passage and direction show movement. They can be described in terms of either two or three dimensions.

Two Dimensions

1. vertical

Examples

over	The plane flew over the mountains.
above	The plane flew above the clouds.
under	The river flows under the bridge.
below	The submarine dived below the polar ice cap.
up	The elevator moved up the shaft.
down	The weakened patient fell down.
off	The specimen slipped off the slide.

2. horizontal

Examples

on	A train moves on rails.
along	The plane flew along the fault line.
up ($\leftrightarrow$)	The submarine moved up the river.
down ($\leftrightarrow$)	The lumber floated down the river.
toward	The ions move toward the cathode.
by (to)	The doctor drove by the hospital to see her patient.
by (past)	The doctor drove by the accident but did not stop.
past	The mud flowed past the village.

3. planar

Examples

(a)round	The earth moves (a)round the sun.
across	The Na$^+$ ions pass across the membrane.
beyond	The probe has now flown beyond Pluto.
through	They watched the reaction through the glass.

EXERCISE 3.5   Add the appropriate preposition of two-dimensional passage or direction to the blanks. There is one sentence for each preposition in the list below.

Example: The plane flew _____ the mountains.
  Answer: The plane flew over the mountains.

above	beyond	off	toward
across	by (past)	on	under
along	by (to)	over	up
(a)round	down	past	up ($\leftrightarrow$)
below	down ($\leftrightarrow$)	through	

1. Oxygen is required when flying ——————— a certain altitude.
2. Navigational instruments prevent aircraft from flying ———————— course.
3. Dr. Wesson had to stop ——————— the lab to pick up the computer printout.
4. No one is allowed ——————— this point because of radiation.
5. Flowmeters measure how much liquid flows ————— a certain point.
6. Heavy rains caused water to flow ——————————— the dam.
7. Ants move ———————————————————— a scent trail.
8. Cruise missiles fly ————————————————— the level of radar.
9. The water flowed ————————— the street, flooding several stores.
10. The geological team lowered a man ————————— into a hole in the glacier to examine the stratified snow layers.
11. The passage of blood can be heard as it moves ——————— the heart valves.
12. The Voyager probe took highly detailed color photographs as it flew ————————————— Jupiter and Saturn.
13. Water seeping ——————— a dam can dangerously weaken the foundation.
14. The galaxy rotates ————————— its center, which some astronomers believe to be a black hole.
15. The explorers traveled ————————— the river to discover its source.
16. Pelicans skim ————————— the surface of the sea in search of fish.
17. Plants usually grow ————————————————— a source of light.
18. Snow tires are required when driving ————————— mountain roads in winter.
19. Liquids move ——————— the chromotography paper to characteristic levels.

Three Dimensions

    1. open

            Examples

across   The comet streaked across the night sky.
through   The bullet passed through the heart.
around   Anemones move around the ocean floor. (I.e., they
                move randomly and continuously.)

    2. closed

            Examples

into   The pharmacist poured the solution into a beaker.
out of   The young bird broke out of its shell.

all over	The airline flies all over the world.
throughout	The cancer has spread throughout the patient's lungs.
around	Electrons move around the nucleus. (I.e., they move in a roughly spherical pattern.)

EXERCISE 3.6   Add the appropriate preposition of three-dimensional passage or direction to the blanks. There is one sentence for each preposition in the list below.

Example: The water is poured _____ a flask.
Answer: The water is poured into a flask.

| across | around | out of | throughout |
| all over | into | through | |

1. The damaged ship was lifted_____ the water.
2. The pressure in the flask caused the stopper to fly _____ the room.
3. Some cosmic rays pass right _____ the earth and out the other side.
4. Light moves at the same speed _____ the universe.
5. After partial digestion in the stomach, food passes _____ the duodenum.
6. The victim sustained second-degree burns _____ his body.
7. Acoustic insulation prevents sound from bouncing _____ a concert hall and becoming distorted.

THE PREPOSITIONS *AS* AND *LIKE*

The words *as* and *like* have many different functions in English. However, in this section, we will consider only the prepositional uses of these words.

as

The preposition *as* generally means *equal to*. Look at these examples:

The heart functions <u>as</u> a pump for the blood.
   (the heart = a pump)
Dr. Calo worked <u>as</u> a research biologist for several years.
   (Dr. Calo = a research biologist)

like

The preposition *like* generally means *similar to*. Look at these examples:

The heart functions <u>like</u> a mechanical piston pump.
   (the heart ≠ mechanical piston pump)

The pineal gland is <u>like</u> a small pine cone.
(the pineal gland ≠ a small pine cone)

as versus like

<u>As</u> a nurse, Mary assisted the surgeon in several operations. (Mary is a nurse.)
<u>Like</u> a nurse, Anna's mother took care of her until she recovered. (Anna's mother is not a nurse, but she worked in the same way as a nurse.)

Incorrect:
*<u>As</u> all children, young birds depend on their parents for food. (they ≠ all children)
*There are several means of crossing a bay, <u>as</u> a bridge, a ferry, and a tunnel.

Correct:
<u>Like</u> all children, young birds depend on their parents for food.
There are several means of crossing a bay, <u>such as</u> a bridge, a ferry, and a tunnel (see Unit IV, Section 2, "Listing").

EXERCISE 3.7　　Add *as*, *like*, or *such as* to the following sentences.

Example: Termites, _____ other social insects, live in colonies.
Answer: Termites, like other social insects, live in colonies.

1. The nose functions _____ both a filter and a warmer of incoming air.
2. _____ Mars, the planet Neptune has two satellites.
3. Galileo worked _____ a mathematics professor before the invention of the telescope enabled him to make his famous discovery.
4. A glacier acts _____ both a file and a plow at the same time.
5. _____ a larva, an insect consumes more food than at any other stage.
6. Jet propulsion is achieved by means of jet engines, _____ the turbojet, the ram jet, and the pulse jet.
7. A bimetallic strip functions _____ a thermometer in certain applications.
8. The sun, _____ other stars, was formed from a cloud of hydrogen gas.
9. Many modern appliances, _____ dishwashers, consume a lot of energy.
10. The first graph is _____ the second except for the slope of the plotted line.

ADVERBIAL *AS*

One of the uses of adverbial *as* in scientific statements is to indicate a complete idea that has been mentioned in another sentence. In this case, the subject is usually deleted. Look at this example:

> The velocity reaches a maximum at 39 seconds, <u>as</u> shown in Figure 1.

This kind of sentence is the result of sentence combining and reduction:

> The velocity reaches a maximum at 39 seconds.
> (THIS FACT = the velocity reaches a maximum at 39 seconds.)
> +   <u>AS</u>
> ‾‾‾
> THIS FACT is shown in Figure 1.
> _____
> The velocity reaches a maximum at 39 seconds, as is shown in Figure 1.

Such a sentence is usually reduced further by removing *is:*

> The velocity reaches a maximum at 39 seconds, as shown in Figure 1.

As is used in this way to refer to related information that occurs either before or after the time indicated in the main sentence. This information is often within the text being read and can be in the active or passive form.

> <u>As</u> Rutherford has demonstrated, particles do not always deflect in this way.
> <u>As</u> has been demonstrated, particles do not always deflect in this way.
>
> The enzyme must be manufactured in the body, <u>as</u> we will discuss in Chapter 7.
> The enzyme must be manufactured in the body, <u>as</u> will be discussed in Chapter 7.
>
> <u>As</u> Pasteur proved, bacteria in milk can be destroyed by heat.
> <u>As</u> was proved in 1857, bacteria in milk can be destroyed by heat.

Adverbial *as* is sometimes confused with the preposition *like:*

> Incorrect:
> *<u>As</u> Tokyo, Saõ Paulo has a problem with automobile pollution.
> (Tokyo is not Saõ Paulo.)
> *<u>Like</u> in Tokyo, Saõ Paulo has a problem with automobile pollution.
> (This is considered substandard English.)

Correct:

As in Tokyo, Saō Paulo has a problem with automobile pollution. (The problem of automobile pollution exists in both Tokyo and Saō Paulo.)

Like Tokyo, Saō Paulo has a problem with automobile pollution. (Tokyo is similar to Saō Paulo in that both have automobile pollution problems.)

EXERCISE 3.8 Fill the blanks with *as* or *like*.

Example: ATP is generated at three sites, _____ shown in Figure 14-4.
 Answer: ATP is generated at three sites, as shown in Figure 14-4.

1. The cost of an oil furnace is much lower, _____ is shown in the appendix.
2. A weather forecast is only approximate, _____ all complex predictions.
3. _____ Roentgen demonstrated in 1895, X rays can make a photographic image of the interior of a solid.
4. _____ Roentgen, Max von Lau was interested in the application of X rays.
5. Cases _____ this should be referred to a specialist.
6. The chemical elements are arranged according to periodic law, _____ was determined by Mendeleev in 1869.
7. The amount of contaminated blood necessary for the transmission of viral hepatitis, Type B, is microscopic, _____ evidenced by the report of infection in five of ten volunteers injected with 0.00004 ml of whole blood.
8. Chemists find it convenient to picture an electron _____ a blurred photograph of rapid movement, _____ a cloud around the nucleus.
9. It appears that the earth's core is not a nickel-iron alloy, _____ had previously been believed, but a mixture of iron and some lighter elements, _____ oxygen and sulfur, which have metallic properties at high pressures.
10. At $-455°$ F, materials _____ niobium become superconductors and lose virtually all electrical resistance.

Section 4
VERBS

SEQUENCE OF TENSES

When a verb occurs in either an adverbial clause of time or a noun clause, the verb in the main sentence acts like a "magnet" on the verb(s) in the clause, "pulling" the clause verb into the same general tense. This effect is called sequence of tenses. Look at these examples:

> Adverbial clause of time:
> High tides OCCUR when the oceans <u>are drawn</u> toward the moon.
> Little WAS KNOWN about microorganisms before Jansen <u>invented</u> the microscope.
> Noun clause:
> The ancient Greeks WERE aware that rubbing amber <u>would electrify</u> it.
> The report STATED that the elephant-seal population <u>was increasing</u>.
> Adverbial clause and noun clause combined:
> Edison TESTED many heat-resistant materials until he <u>discovered</u> that a simple cotton thread <u>was</u> an excellent filament for an electric light bulb.

The basic "magnetic" effect of the main verb is to keep the tense in the same general classification of past or present. Look at this example: The weatherman says that it will rain.

FORM	PAST	PRESENT
Simple	The weatherman said that it always rained.	The weatherman says that it always rains.
Continuous	The weatherman said that it was raining.	The weatherman says that it is raining.
Perfect	The weatherman said that it had rained.	The weatherman says that it has rained.
Modal	The weatherman said that it would rain.	The weatherman says that it will rain.

In other words, if the main verb is in one of the past tenses (simple past, past continuous, past perfect, past modal), then the clause verb must also be in one of the past tenses. If the main verb is in one of the present tenses (simple present, present continuous, present perfect, present modal), then the clause verb must be in one of the present tenses. Only when an adverbial phrase contains the word *since* can a past verb be used when the main verb is in a present tense:

> The computer HAS WORKED perfectly since it <u>was repaired</u> last month.
> The universe HAS BEEN EXPANDING since it <u>began</u> explosively 15 billion years ago.

This is because *since* shows the effect of a past event on a present situation (see Unit IV, Section 4, "The Present Perfect Tense").

EXERCISE 4.1   Put the verb in parentheses into the correct form. Underline the controlling verb.

Example: The continents formed when the earth (be) _____ young.
 Answer: The continents <u>formed</u> when the earth was young.

1. An earthquake occurs when sufficient strain (develop) _____ between adjacent rock masses.
2. Berzelius, a Swedish chemist, discovered that when selenium (place) _____ in light, it (conduct) _____ electricity.
3. Solar-plant engineers hope that in ten years photovoltaic technology (be) _____ cost-efficient.
4. A fire burns until all combustible material (consume) _____.
5. Synthetic plastics have replaced many traditional materials since the phenol plastic Bakelite (develop) _____ in 1909.
6. Henry Fox Talbot, an English chemist, demonstrated that certain elements (produce) _____ certain dark lines in a spectrum.
7. Fleming discovered penicillin while he (work) _____ with the staphylococcus bacteria.
8. The process of electroplating is based on the principle that certain liquids (ionize) _____ when an electric current (pass) _____ through them.
9. In 1610, Galileo proved that Copernicus (be) _____ correct when he (state) _____ in 1543 that the earth (move) _____ on its own axis.
10. Congestive heart failure is a syndrome that (result) _____ when the ventricles (fail) _____ to pump sufficient blood to meet the body's needs.

SEQUENCE OF TENSES WITH FACTS

An exception to the sequence-of-tenses rule occurs with noun clauses in scientific writing. If the clause expresses a fact or an activity that is relatively permanent, it is written in the simple present tense (like most facts), even if the main verb is in the past:

Magellan PROVED that the earth <u>is</u> round.
Watson and Crick DISCOVERED that the DNA molecule <u>has</u> the shape of a double helix.

In this case, the past main verb cannot be a verb that expresses simply a personal idea (e.g., *thought, believed, guessed*).

Incorrect:
*Magellan thought that the earth is round.
*Magellan believed that the earth is round.
Correct:
Magellan THOUGHT that the earth was round.
Magellan BELIEVED that the earth was round.

EXERCISE 4.2 Put the verb in parentheses in the correct form. Underline the main verb.

Example: Lavoisier proved that the metals (be) _____ elements.
Answer: Lavoisier proved that the metals are elements.

1. Theoretical science began when the Greeks (start) _____ to ask questions about what things (make) _____ of and where they (come) _____ from.
2. Ptolemy believed that the sun (revolve) _____ around the earth.
3. Ptolemy did not realize that the earth (revolve) _____ around the sun.
4. Vesalius named every bone, every muscle, and most of the blood vessels in the human body, but he did not know how the body (function) _____ .
5. Johannes Kepler was the first to show how a planet (move) _____ .
6. Before the discovery of oxygen, many scientists thought that when a substance (burn) _____ , its "phlogiston" (escape) _____ into the air.
7. Jules Verne, a nineteenth-century science-fiction writer, imagined that a rocketship (can fly) _____ from the earth to the moon.
8. When Mendel's 1866 report (rediscover) _____ in 1900, biologists found that Mendel (make) _____ many important discoveries concerning heredity.
9. Christiaan Eijkman demonstrated that a shortage of vitamin B_1 over a long time (can cause) _____ the nerve disease called beriberi.
10. Hideki Yukawa proposed a theory that (account for) _____ the type and magnitude of forces that (hold) _____ the atomic nucleus together.

Section 5
WRITING AIDS

WORDINESS

Wordiness is the use of too many words to express an idea. Scientific and technical writing must be clear and concise; wordiness makes writing weak and vague. Two common forms of wordiness are (1) indirect contstructions and (2) weak verbs.

Indirect Constructions

Indirect constructions are primarily sentences beginning with *it* or *there*. Sentences beginning with *there* (the unstressed form) are common in English because they tell us that something exists. Such sentences are wordy only if they are immediately followed by a sentence or clause that states **what** exists. In this case, the two should be combined.

> Incorrect:
>   There are three forms of water. They are the solid, liquid, and gaseous phases. (fourteen words)
>   There is a famous equation that was discovered by Einstein. It is $E = mc^2$. (thirteen words)
> Correct:
>   The three forms of water are the solid, liquid, and gaseous phases. (twelve words)
>   The famous equation discovered by Einstein is $E = mc^2$. (eight words)

Sentences beginning with *it* are also common in English. They too are wordy only if they are immediately followed by a sentence or clause that should have been in the subject position.

> Incorrect:
>   It is with the third process that this paper is concerned. (eleven words)
>   It seems that the new unified field theory is better than the old one. (fourteen words)
> Correct:
>   This paper concerns the third process. (six words)
>   The new unified field theory seems better than the old one. (eleven words)

EXERCISE 5.1   Correct the following wordy sentences.

Example: There are two kinds of bridges; they are stationary and movable.
Answers: The two kinds of bridges are stationary and movable.
           Bridges are either stationary or movable.

1. It was shown by Mendel that many characteristics are inherited as separate units.
2. There is one active ingredient in aspirin: acetylsalicylic acid.
3. Reports stated that there were 200,000 people killed by a cyclone-driven tidal wave from the Bay of Bengal.
4. It is fire that all animals except domesticated dogs and cats are afraid of.
5. There is an average of 5 million red blood cells in the human body.
6. It is from the words "quasi-stellar object" that the term *quasar* is derived.
7. It has been proved that plutonium is a carcinogen.
8. There is a principle that was first stated by Bernoulli; it is that at any point in a fluid, under conditions of steady flow, the amount of energy per unit mass is constant.
9. It was the fact that the chemicals were impure that caused a delay in the first test.
10. There is evidence that tends to confirm that the universe began its existence about 15 million years ago.

Weak Verbs

A verb is weakened when it is made into a noun phrase, thus requiring the addition of another verb. Such a structure is considered wordy because it uses more words to say the same thing. Look at this example:

The doctor <u>performed an examination</u> of the patient. (eight words)

In this sentence, the weak verb *performed* has been added because the verb *examine* was made into the noun *examination*. To correct this sentence, we remove the additional verb and change the noun back into its original verb form.

The doctor <u>examined</u> the patient. (five words)

Here are some further examples:

Incorrect:
  The graduate students <u>carried out experiments</u> on white rats. (nine words)
  <u>Analyses were made</u> of each sample. (six words)

Correct:
>The graduate students <u>experimented</u> on white rats. (seven words)
>Each sample <u>was analyzed</u>. (four words)

EXERCISE 5.2   Correct the following wordy sentences.

Example: The engineers made a design of a new oil pump.
 Answer: The engineers designed a new oil pump.

1. The temperature change will produce an increase in vapor pressure.
2. A transistor stage provides amplification of the radio signal.
3. Dr. Gomez conducted an analysis of the geological structure of the Andes.
4. A hospital must furnish transportation for emergency patients.
5. The size of the cyclotron presents serious limits to the size of the beam.
6. The installation of the 200-inch reflector was completed in early March.
7. The exposure of the photographic plate gave confirmation of the radioactive nature of uranium.
8. The fuse mechanism brings about automatic disconnection if a short occurs in the electrical system.
9. The deterioration of the alpha wave occurs rapidly.
10. New techniques have improved the survival rate by almost double.

Another form of weak-verb wordiness occurs when we weaken a strong verb by making it into an infinitive or V_{ing} form, which requires the addition of another verb (see also Unit IV, Section 4, "Verbs With Inanimate Subjects") or an adjective. Look at these examples:

>A thermometer <u>is used to measure</u> temperature.
>Galileo <u>was correct in proving</u> that the solar system is heliocentric.

To correct the first sentence, we remove the weak verb *is used* and change the infinitive back to a finite verb. To correct the second sentence, we remove the adjective *correct* with its allied preposition *in* and change the V_{ing} form *proving* back to a finite verb:

>A thermometer <u>measures</u> temperature.
>Galileo <u>proved</u> that the solar system was heliocentric.

Here are some further examples:

Incorrect:
>Methyl-ethyl ketone <u>causes</u> the oil <u>to dissolve</u>.
>The programmer <u>succeeded in debugging</u> the program.

Correct:
    Methyl-ethyl ketone <u>dissolves</u> the oil.
    The programmer <u>debugged</u> the program.

EXERCISE 5.3   Correct the following wordy sentences.

Example: A thermometer is used to measure temperature.
 Answer: A thermometer measures temperature.

1. The compiler is used to translate programming languages into machine language.
2. Researchers have been successful in establishing the relative strength of known carcinogens.
3. The catalyst proceeds to initiate the reaction.
4. The ozone layer of the atmosphere acts to reduce the amount of ultraviolet radiation that strikes the earth.
5. A doctor is limited to having only two alternatives in such a case.
6. A venturi meter serves to monitor the change in flow.
7. The entire atmosphere is known to weigh 5,700,000,000,000,000 (5.7×10^{15}) tons.
8. An irregular heartbeat pattern causes an alarm to sound.
9. A muffler is employed to reduce the noise of an internal-combustion engine.
10. The hypothalamus operates to coordinate body-temperature control.

EXERCISE 5.4   Correct the wordiness problems in the following paragraph.

A NEW FOOD PRESERVATION PROCESS[2]

There is a new food preservation process that is substantially more energy-efficient than canning or freezing. It has been developed by University of Maryland scientists. It is named Gaspak, and the new process requires putting a food into a chamber and removing all air. Subsequently, the food is subjected to treatment with a combination of gases. They are carbon monoxide, sulfur dioxide, and other gases. This is to inhibit bacterial growth and deterioration. The treated food is then packaged in a germ-free container which is filled with the gas.

It is a new method which functions to permit food to be kept fresh at room temperature in many places. These include warehouses, stores, and homes. Tests served to indicate that food samples were well-preserved and palatable after 30 days and after 250 days.

[2]Adapted from *Information Please Almanac 1981.* Copyright © 1980 by Houghton Mifflin Company. Reprinted by permission of Houghton Mifflin Company.

KEY-PHRASING

Key-phrasing is an internal method of cohesion in paragraphs. By means of key-phrasing, "important information given in one sentence is referred to again in a later sentence."[3]

Key-phrasing usually makes use of the word *this* plus a noun phrase. We must be careful to use the word *this* and not *that*. *That*, related to the word *there* (the stressed form), causes the reader to look outside the paper. *This*, related to the word *here*, directs the reader back to what was just said.

Key-Phrasing with Repeated Words or Synonyms

The simplest form of key-phrasing is the use of the word *this* (pl. *these*) in four situations: (1) by itself, (2) with a repeated word, (3) with part of an earlier noun phrase, or (4) with a synonym. Look at these examples:

(1) *this* by itself:
Potassium increases the strength of plant tissues. This helps the plant to withstand mechanical damage.
(2) *this* + a repeated word:
Many processed foods contain additives. These additives generally retard spoilage and replace vitamins lost in cooking.
(3) *this* + part of an earlier noun phrase:
Alzheimer's disease causes senility in older people. This disease may be caused partly by aluminum in the diet.
(4) *this* + a synonym of an earlier noun phrase:
a. The sensory and motor-nerve centers make up the major portion of the brain. This organ is situated in the cranium.
b. Scientists have discovered a body of magma under much of the eastern U.S. This pocket of molten rock may one day be used as a heat source for generating electrical power.

In sentence 1, *this* refers to the entire sentence rather than to the subject or the object alone; the strength of plant tissues resulting from potassium helps the plant to withstand mechanical damage (see also "Adverbial *As*," Section 3 of this unit.) Such a structure occurs in spoken English a great deal but is considered by some to be not formal enough for scientific writing.

As you can see in sentence 4a, the synonym of an earlier noun phrase is usually a generalized word to which the phrase belongs (the brain belongs to the group of things called organs). However, as you can see in sentence 4b, a synonym in this situation sometimes gives us additional information about the earlier noun phrase (a body of magma is a pocket of molten rock). This is called an implied definition (see also Unit IV, Section 3, "Appositives").

---

[3]John Swales, *Writing Scientific English* (London: Thomas Nelson and Sons, 1979), p. 109. The term *key-phrasing* comes from this book.

EXERCISE 5.5   Fill the blank in the second sentence with *this* or *these* plus an appropriate word or phrase from the first sentence.

Example: The first good roads were built by the Romans. ————————————
were mainly for military use.
Answer: The first good roads were built by the Romans. These roads were mainly for military use.

1. Suspension bridges depend upon giant wire cables for support. ———————————— are usually manufactured at the site.
2. A mucous membrane covers the tongue. ——————————— contains four types of tiny bumps, or *papillae*.
3. Air rising over a moist region may cause cumulus clouds to form in cooler air above the surface. ——————————— darken to rain clouds as they receive more moisture.
4. Each spiral nebula seems to be rotating and throwing off matter at its rim. ——————————— may condense into stars.
5. The autonomic nerves are either motor or sensory. ————————— are further divided into sympathetic and parasympathetic.
6. The volume of water in the oceans is fourteen times that of all land above sea level. ——————————— , frozen into a ball, would form a globe 850 miles in diameter.
7. The generator design can be suggested by three loops. ————————— ————————— are mounted on a common shaft 120 degrees apart.
8. Small buildings often contain elevators that passengers operate themselves. In ——————————— , the doors usually open and close automatically.
9. The atmospheric pressure at sea level is the standard pressure for purposes of weather forecasting. ——————————— , under normal conditions, will support a column of mercury 29.92 inches high.
10. A huge amount of water still remains locked in the ice over Antarctica and Greenland. If ——————————— melted, it would raise the oceans about 200 feet.

EXERCISE 5.6   Fill the blanks with *this* or *these* plus the appropriate word or phrase from the list below.

Example: Carbon is present in all living matter. ——————————— occurs in its pure form as diamond or graphite.
Answer: Carbon is present in all living matter. This element occurs in its pure form as diamond or graphite.

basis	element	instrument	random activity
condition	evidence	material	
defenses	fact	phenomenon	

1. During his experiment, Hertz found that light falling upon metal would drive out a negative charge. ———————————— is called the photoelectric effect.
2. The porcupine is noted for its armor of quills or spines. ———— usually lie back on its body but are raised in time of danger.
3. Liquids in a vacuum boil at a temperature lower than the normal boiling temperature. ———————————— has been useful to food and pharmaceutical manufacturers.
4. In the sixteenth century, Paracelsus said that a healthy body required salt, sulfur, and mercury in the proper proportions. On ————————————, he practiced medicine and attracted many followers.
5. A human infant's body is in constant motion. ———————————— is very tiring for the baby.
6. Ores contain earthy materials such as sand and clay. ———————— are worthless and are separated from the metal and discarded.
7. In the infrared spectral recorder, a photoelectric cell translates weak infrared rays into electric current. ———————————— helps scientists to study conditions on distant planets.
8. Rocks that contain the remains of sea creatures are found in plains and mountains. ———————————— proves that the earth's surface moves both downward and upward.
9. In group VIIA of the periodic table are the halogens: fluorine, chlorine, bromine, iodine, and astatine. ———————————— are all nonmetallic.
10. Corn will grow wherever there is suitable soil, ample soil moisture, freedom from cold and frost, and hot sun. ———————————— are found around the Mediterranean, in India, in South Africa, and in the entire Western Hemisphere.

Key-Phrasing with Altered Forms of Earlier Words

The more important form of key-phrasing is the use of a **different form** of an earlier word. Often, the first occurrence is a verb structure and the second occurrence is a noun structure, usually with *this*. Look at these examples:

Faraday <u>discovered</u> the principle of the induction coil in 1831. <u>This discovery</u> lead to the development of motors and generators.

In the Haber-Bosch process, nitrogen <u>reacts</u> with hydrogen in the presence of an iron catalyst to produce ammonia, NH_3. <u>This reaction</u> is the most widely used industrial method of nitrogen fixation.

It is also possible for the first occurrence to be an adjective structure:

> Waves of <u>different</u> lengths can travel on the same lines without mixing. <u>This difference</u> enables numerous telephone conversations to be transmitted over the same wire.

EXERCISE 5.7   Fill the blanks in the second sentence with *this* or *these* plus a noun derived from a word in the first sentence.

Example: Steel plates are sometimes bonded with stainless steel or other metals. _____ prevents food or other products from reacting with the steel.

Answer: Steel plates are sometimes bonded with stainless steel or other metals. This bonding prevents food or other products from reacting with the steel.

1. Germs can irritate the nasal membranes. _____ causes the membranes to secrete a watery liquid to wash away the germs.
2. If a mixture of two liquids is heated, the more volatile liquid will vaporize at a lower boiling point. When _____ is passed through a condenser, it is changed back to a liquid.
3. A chemical atom rearranges the electrons surrounding the nucleus to a more stable form. _____ can result in a sharing of atoms between electrons.
4. When it rains, some of the rain runs off over the ground surface. _____ becomes a stream and eventually a river.
5. A family of organic compounds has similar properties. _____ arises from the properties of some characteristic group in each family.
6. An earthquake is produced when one tectonic plate moves against another. _____ is recorded on a seismograph.
7. Diamond is the hardest substance known. _____ gives a cutting edge of almost endless wearing quality.
8. The amount of solar energy received by a region depends on the combined effect of distance from the equator and the seasonal factor. _____ gives the region what is called its solar climate.
9. Worker bees dance to inform the others in the hive about sources of nectar. _____ indicates the distance of the source from the hive and the angle of the source in relation to the sun.
10. Compasses on a ship are affected by the magnetic force of the ship itself. The effect of _____ is called deviation, and it is used to adjust the compass error.

PART II
Writing an Abstract

Section 6
PREWRITING ACTIVITY: SUMMARIZING

An abstract is a special kind of summary. It is used in scientific writing to summarize the major content of a report or study. Before we discuss the form of a scientific abstract, we need to practice writing a general summary.

EXERCISE 6.1   Summarize one of the models in Section 8 of Unit II, III, or IV.

Examples: Latex (Unit I, Section 8, Model 2)
    Latex is a natural substance found in rubber trees. Heated with carbon and sulfur, latex forms commercial rubber, which is used primarily to make truck tires.

The Canine Teeth (Unit I, Section 8, Model 3)
    The canine teeth (from the Latin word for *dog*) are the strongest and most decay-resistant teeth in the mouth. In humans, they guide the teeth in chewing and are commonly used as anchors for dental work. In animals, they are used for tearing food and for defense.

Section 7
STRUCTURE

An abstract is usually the first contact that a prospective reader has with a technical report. In fact, the reader normally uses the abstract to decide whether or not the report is worth reading. For this reason, the abstract must be written with a clear idea of the intended audience, it must attract the reader's attention, and it must be written in simple, nontechnical terms as much as possible. Because we need a thorough understanding of the material to be summarized before we can write an abstract, we usually write it after we have completed the whole report.

Abstracts can be descriptive, evaluative, or informative. A typical scientific abstract is informative. It presents the objective, states the results and conclusions, and either makes recommendations or discusses the implications of the report to which it is attached. The abstract is usually about six to eight lines long, representing 2 to 5 percent of the total paper, and it usually has the form of a paragraph. However, different publications have their own styles, and the abstract is sometimes broken into short paragraphs.

PART 1: OBJECTIVE

The first sentence of an abstract states why a project was carried out and/or why the report was written. (This sentence by itself is often called a descriptive abstract; it is sometimes found below the title of a scientific paper, as in *Scientific American*.) Examples:

> The formation of nitrogen compounds in engine exhaust gas was investigated.
> Cis-trans isomerism in straight-chain, internal olefins was determined.

The materials and methods used in a study are usually **not** focused upon in an abstract, but they can be included in the first sentence.

> "Time-lapse photography was used to study cell-cycle progession and cell-division abnormalities in rat 9L brain tumor cells after treatment in culture with 2 g/ml and 5 g/ml of 1,3 bis (2-chloroethyl)-1-nitrosourea (BCNU)."[4]

PART 2: RESULTS AND CONCLUSIONS

The second sentence (or part) of an informative abstract describes the actual results and/or conclusions of the report. Examples:

> Principal localizations of the tumors were the nose, eyelids, lips, and ears, with 8, 7, 4, and 2 localizations, respectively.
> Results show that the rate of particle deposition agrees with experimental data.
> Symptoms of X ray damage occurred in far more exaggerated forms than those displayed by other mammalian cell types.

[4]Reprinted with permission from the *European Journal of Cancer, 15*, U. K. Ehmann and K. T. Wheeler, "Cinemicrographic Determination of Cell Progression and Division Abnormalities After Treatment With 1,3 Bis (2-Chloroethyl)-1-Nitrosourea," Copyright 1979, Pergamon Press, Ltd.

PART 3: IMPLICATIONS AND RECOMMENDATIONS

The third sentence (or part) of an abstract discusses the implications of the findings and/or makes recommendations based on those findings. Examples:

> The low survival of L5178YS/S cells after irradiation may be related more to faulty cytokinases than to chromosomal aberrations.
> The chemical composition of particulates should be further studied in relation to exhaust-gas temperature.
> The combined use of statistical and quantitative analyses is recommended when characterizing urban roadway dust.

Section 8
MODELS

MODEL 1: INSECT AND RODENT CONTROL[5]

Descriptive Abstract

Ten widespread diseases that are hazards in isolated construction camps can be prevented by removing or destroying the breeding places of flies, mosquitoes, and rats, and by killing their adult forms.

Informative Abstract

Ten widespread diseases that are hazards in isolated construction camps can be prevented by removing or destroying the breeding places of flies, mosquitoes, and rats, and by killing their adult forms. The breeding of flies is controlled by proper disposal of decaying organic matter, and of mosquitoes by destroying or draining pools, or spraying them with oil. For rats, only the indirect methods of rat-resistant houses and protected food supplies are valuable. Control of adult forms of both insects and rodents requires uses of poisons. Screens are used for insects. Minnows can be planted to eat mosquito larvae.

Full Report

Introduction. Flies, mosquitoes, and rats are the vehicles of infection for ten widespread diseases. Flies, which are mechanical carriers, are responsible for the transmission of the intestinal diseases; i.e., (1) typhoid, (2) paratyphoid,

---

[5]Jerry Garrett, "Sanitation Requirements for an Isolated Construction Project," as cited in Gordon H. Mills and John A. Walter, *Technical Writing* 3rd. ed. (New York: Holt, Rinehart & Winston, 1970), pp. 96–98.

(3) dysentery, (4) cholera, and (5) hookworms. Mosquitoes spread diseases by biting; they are vectors in the cycle of transmission of (6) malaria, (7) yellow fever, and (8) dengue. Rats are the reservoirs of (9) plague and (10) typhus, but the rat's fleas are the vehicles of transmission.

There is but one way to stop the spread of these diseases, and that is to get rid of the insects and rodents, and the most effective way of getting rid of them is to remove their breeding places by good general sanitation. The only alternative is to kill the adults. Positive steps which may be taken in these operations are discussed below.

Breeding control. As pointed out above, if there are no insects or rodents, the diseases which depend on them for transmission must vanish. It is certainly cheaper and simpler to destroy their breeding places than to try to kill billions of adults only to find more billions waiting to be killed.

Flies. One characteristic of the fly makes it particularly susceptible to breeding control. The fly always lays its eggs in decaying organic matter, preferably excreta or manure. Three stages in the life of the fly—the egg, larva, pupa—are spent in the manure. A minimum of eight to ten days is spent here before the adult emerges. Therefore, the measures are relatively simple. First, there should be proper sewage disposal, i.e. the flies are never permitted to come into contact with human excreta. Secondly, all animal manure should be removed within four or five days, or in other words, before pupation takes place. The manure should either be placed in fly-proof storage bins or tightly compressed so that the adult fly cannot emerge after pupation. The final breeding control is to destroy all decaying organic matter such as garbage by either burying it two feet deep or burning it.

Mosquitoes. It is not as simple to control the breeding places of the mosquito as it is to control those of the fly. But it can be done! First, it must be realized that there are many kinds of mosquitoes and that only a few are disease vectors. Still they must all be killed to be sure the correct ones are dead, and they are all important as pests anyway. The female *Aedes aegypti* is the vector for yellow fever and dengue; this mosquito breeds only in clean water in artificial containers. In the southern section of the United States (the chief malaria area in the United States), the malaria vector is the *Anopheles quadrimaculatus*, a night biter, which breeds in natural places, particularly where the water is stationary and where there is vegetation and floating matter to protect the eggs, larvae, and pupae.

Therefore, the best way to prevent the breeding of mosquitoes is to remove all water in which they breed by draining or filling pools, and removing or covering artificial containers. However, since the construction project is only temporary, the operators will be interested in the most economical measures rather than the most permanent. Artificial containers must still be covered, but it might be cheaper to spread a film of oil over all the natural stationary water rather than to try to drain it or fill in the low spots.

Rats. There are no direct ways to control the breeding of rats or their fleas, but sufficient control can be exerted to make them take their breeding elsewhere. This is done by building rat-resistant houses and by preventing the rats from reaching food.

Adult control. *Flies.* Houses should be screened to keep the flies from getting to food. Then, traps such as the standard conical bait trap should be distributed. The most attractive baits, as established by experiment, are fish scraps, overripe bananas, and a bran and syrup mixture. DDT may be used effectively to leave a residual poison for flies.

Mosquitoes. If a house is well screened, the mosquitoes cannot get into the house to bite their victims. Advantage can be taken of the mosquitoes' natural enemies by stocking waterways with minnows which eat the larvae. Poisons which may be used against mosquitoes are DDT and pyrethrum.

Rats. Besides carrying diseases, the rat of course destroys much property. Usually, however, the construction project operator need be concerned with rats only to the extent that they endanger his workers' health. Poisons which may be used against rats are barium carbonate, red squill, 1080, and antu. Other effective means of getting rid of rats are by trapping and fumigation.

MODEL 2

A specimen of steel was tested to determine whether a job lot owned by Northern Railways could be used as structural steel members for a short-span bridge to be built at Peele Bay in northern Alaska. The sample proved to be G40.12 structural steel, which is a good steel for general construction but subject to brittle failure at very low temperatures.

Although the steel could be used for the bridge, we consider that there is too narrow a safety margin between the $-51°$ C temperature at which failure can occur, and the $-47°$ C minimum temperature occasionally recorded at Peele Bay. A safer choice would be G40.8C structural steel, which has a minimum failure temperature of $-62°$ C.[6]

MODEL 3

The Pacific Missile Test Center (PACMISTESTCEN) was requested by the Naval Air Development Center (NAVAIRDEVCEN) to test and evaluate the speech intelligibility characteristics of the HGU-33/P and HGU-34P/P flight helmets configured with Electrovoice AV-993 1000 impedance earphones.

The EV-993 earphones in the HGU-33/P and HGU-34/P flight helmets met the intelligibility criterion for a military communication system and were equal in intelligibility to the standard APH-6D flight helmet configured with H-87B/U EV-992 earphones.

[6]Ron S. Blicq, *Technically–Write!* 2nd ed., © 1981, p. 146. Reprinted by permission of Prentice-Hall, Inc., Englewood Cliffs, N.J.

Appropriate mating connectors and communications procedures analyses are required to integrate the HGU-33/P and HGU-34/P flight helmets in Naval aircraft.[7]

Section 9
ANALYSIS

The basic structure of an abstract, which we discussed in Section 7, is as follows:

  I. Objective
 II. Results and/or conclusions
III. Recommendations or implications

EXERCISE 9.1   With another student, read the abstract in Model 1. Then read the full report from which the abstract is derived. Underline exactly where each piece of information in the abstract was taken from the full report. Discuss your results with your partner. Do you agree?

EXERCISE 9.2   With another student, preferably from your field, analyze Models 2 and 3. Determine the function (objective, result or conclusion, recommendation or implication) of each sentence in the model. Discuss your results with your partner. Do you agree?

Section 10
CHOOSING A TOPIC

Your assignment is to write an informative abstract of a paper in your own field. The best source of papers for use in this exercise is *Scientific American,* although another journal or even a chapter of a textbook would work as well. In writing your abstract, the following procedure is recommended.

1. Most scientific articles are divided into sections. In older issues of *Scientific American,* these divisions are indicated by a double space between paragraphs. After you have determined which paragraphs of the article belong to each section, write a subtitle for each section (a subtitle is the title for a section). This will establish the overall organization of the article.
2. Using your knowledge of paraphrasing, write a summary sentence for **each** paragraph in the article. Be careful not to alter the meaning or sense of the paragraphs. When you have finished, you should have a list of thirty to forty sentences.

[7]Theodore A. Sherman, Simon S. Johnson, *Modern Technical Writing,* 4th ed., © 1983, p. 221. Reprinted by permission of Prentice-Hall, Inc., Englewood Cliffs, N.J.

3. Determine which sentences are concerned with objectives, results or conclusions, and recommendations or implications, and place a check next to them. Ignore the other sentences, which will be concerned for the most part with detailed methods and materials.
4. From the remaining sentences, determine which six to eight sentences are crucial to the understanding of the entire article. Write these into a single paragraph, adding sentence connectors if necessary to make the paragraph more readable.

EXERCISE 10.1   Write a six-to-eight sentence abstract of an article from *Scientific American* (or some other scientific journal or text that does not have an abstract already), using the procedure above.

UNIT VI
The Research Report and the Feasibility Study

PART I
Grammar

Section 1
ARTICLES

ARTICLES WITH PROPER NOUNS

Proper nouns are names or titles of distinct persons, places, or things. Since they are distinct, proper nouns occur only with the articles *the* and *∅*, never with *a(n)*. As a rule, names require the zero article whereas titles require *the*.

	Name	Title
People	President Carter	the president of the U.S.
	Mr. Churchill	the prime minister of England
	Secretary Schultz	the secretary of state
	Dr. Lau	the dean of the School of Engineering
Countries	Korea	the Republic of Korea
	Russia	the Union of Soviet Socialist Republics (USSR)
States	Maine	the state of Maine
Cities	Tokyo	the city of Tokyo

The difficulty with articles and proper nouns is that the rules concerning proper nouns other than people and political divisions (e.g., countries, states, cities) are quite arbitrary.

Proper Nouns with *the*

Proper nouns require *the* if they refer (1) to certain geographical features (oceans, rivers, canals, deserts, forests, and the plural form of islands, lakes, and mountains) or (2) to certain cultural institutions (associations, commissions, libraries, and museums).

Geographical features.

oceans	the Pacific Ocean
rivers	the Yangtze River
canals	the Panama Canal
deserts	the Sahara Desert
forests	the Black Forest
islands	the Philippine Islands (the Philippines)
lakes	the Great Lakes
mountains	the Himalaya Mountains (the Himalayas)

Cultural institutions.

associations	the American Medical Association (AMA)
commissions	the Nuclear Regulatory Commission (NRC)
libraries	the Boston Public Library
museums	the British Museum

Proper Nouns with Ø

Proper nouns require the zero article if they refer (1) to certain geographical features (areas, continents, valleys, and the singular form of islands, lakes, and mountains) or (2) to certain cultural features (holidays, parks, and streets).

Geographical features.

areas	North Africa, Southern California
continents	Asia, South America
valleys	Desolation Valley

Cultural institutions.

holidays	Christmas
parks	Hyde Park
streets	Blake Street

Proper Nouns with Both *the* and *∅*

A few cultural institutions (buildings, businesses, and universities) occur with *the* or *∅*. The zero article is usually used when the proper-noun phrase begins with a family name.

	Example with *the*	Example with *∅*
buildings	the World Trade Center	Wheeler Auditorium
businesses	the Shell Oil Company	Hewlitt-Packard
universities	the University of Ohio	McGill University

EXERCISE 1.1 Fill the blanks with *the* or *∅*.

Example: _____ Amazon River empties into _____ Atlantic Ocean.
Answer: The Amazon River empties into the Atlantic Ocean.

1. _____ Sears Tower in _____ city of Chicago is one of the tallest buildings in _____ North America.
2. _____ Suez Canal connects _____ Port Said on _____ Mediterranean Sea and _____ Port Taufiq on _____ Gulf of Suez in _____ Red Sea.
3. _____ Little Colorado River begins at _____ Zuni Reservoir in _____ Arizona, just south of _____ Petrified Forest, an extensive exhibit of petrified wood. It flows north of _____ city of Flagstaff and empties into _____ Colorado River, which passes through _____ Grand Canyon National Park.
4. _____ Mount Kilimanjaro (19,340 ft.) is situated in _____ northern Tanzania between _____ Lake Victoria and _____ Indian Ocean. _____ Serengeti National Park lies to the west of it, _____ Masai Steppe to the south, and _____ Yatta Plateau to the northeast.
5. _____ LeConte Hall houses _____ physics department at _____ University of California at _____ Berkeley.
6. Bones from the largest known mammal, the baluchitherium, were found in _____ Gobi Desert in _____ People's Republic of Mongolia. This desert is southeast of _____ Khangai Mountains.
7. _____ Library of Congress, _____ Air and Space Museum, _____ NASA (_____ National Aeronautics and Space Administration), and _____ Department of Agriculture are all

located on _____ Independence Avenue in _____ Washington, D.C.

8. The largest of _____ Hawaiian Islands is _____ Hawaii, also known as _____ Big Island. _____ Kilauea, one of _____ world's most active volcanoes, is located there.

9. When sailing from _____ city of Vancouver, one passes through _____ San Juan Islands to _____ Victoria, the capital of _____ British Columbia, on _____ Vancouver Island. _____ Pacific Ocean lies to the west.

10. _____ Apple Computer Company, situated in _____ "Silicon Valley" near _____ Stanford University, competes with _____ IBM and other companies concerned with microelectronics.

SOME IDIOMATIC USAGES OF ARTICLES

There are a few cases in English where article usage is idiomatic. These include words and phrases with *a(n)*, *the,* and *∅.*

Idiomatic Structures with *a(n)*

a few versus *few.* The phrases *a few* and *few* both indicate small quantities. However, with the article *a* the phrase has a positive or neutral sense, whereas with the zero article it has a negative one. The word *only* can occur only with *a few;* the word *so* can occur only with *few* and *little.* Look at these examples:

> The satellite camera took <u>a few</u> excellent shots of Jupiter's moon Io.
> Only <u>a few</u> elements are liquids at room temperature.
> <u>Few</u> animals live at the polar ice cap.
> So <u>few</u> patients survived that the drug was banned.

The same is true of the phrases *a little* and *little:*

> <u>A little</u> alcohol acts as a stimulant to the body.
> There was so <u>little</u> water that the animals died.

However, the negative sense with the *∅* article sometimes produces an overall positive effect.

> <u>Few</u> scientists have influenced scientific thinking like Einstein. (i.e., he was a great scientist.)
> <u>Little</u> time was lost in getting the patient into surgery. (i.e., the patient was moved rapidly.)

A quarter of versus *half.* Most fractions use an *of*-phrase with the article *a(n)* or the number *one* (see also Unit I, Section 1, "*A(n)* versus *One*"):

> <u>a</u> third of the population
> <u>a</u> quarter of the population
> <u>an</u> eighth of the population

Only the fraction *half* is commonly written without *an:*

> half (of) the population

a(n) with *how, so, as, too,* and *no less.* Words such as *how, so, as, too,* and *no less* attract an adjective and displace the article:

> The doctor did not realize <u>how weak a heart</u> the patient had.
> The coil developed <u>so high a voltage</u> that the rods arced immediately (see Unit IV, Section 2, "The SO Group").
> Alcohol does not have <u>as high a boiling point</u> as water (see "Negative Noun Comparisons," Section 3 of this unit).
> The bridge collapsed because it had <u>too weak a superstructure</u>.
> Nuclear power is <u>no less a problem</u> than nuclear weaponry (see also "Comparisons," Section 3 of this unit).

EXERCISE 1.2 Add *a(n)* to the following sentences **where necessary**.

Example: Air pollution decreased in quarter of the nation's cities.
 Answer: Air pollution decreased in a quarter of the nation's cities.

1. Few atoms of plutonium in a human body are enough to trigger cancerous growth.
2. The estimated U.S. share of world science and technology fell from about third in 1967 to less than quarter in 1980.
3. In studying spectroscopy, G. R. Kirchhoff worked with no less man than Robert Bunsen, inventor of the Bunsen burner.
4. The life expectancy of a kangaroo is about half that of a cat.
5. Few people knew how difficult problem the containment of a fusion plasma would be.
6. The child had so high temperature that he had to be immersed in a cold bath.
7. Only few countries have suicide rates of 20 per 100,000 or higher.
8. The Almendra Dam in Spain has almost as great reservoir as the Chirkey Dam in the U.S.S.R.
9. The investigators knew that the reactor had leaked because little radioactive water was found nearby.
10. Depo Provera offered too high risk for the Federal Drug Administration to recommend its use as an injectable contraceptive.

Idiomatic Structures with *the*

Most diseases have formal names with the *0* article:

cancer	diabetes	influenza	diphtheria	dysentery
hepatitis	typhoid	smallpox	tetanus	meningitis

However, a few common diseases can have the article *the:*

(the) flu (influenza)
(the) measles
(the) mumps

Simple ailments have the article *a(n):*

a cold   an upset stomach
a headache   a broken leg

Body parts are usually referred to with a possessive pronoun since they "belong" to each human being:

The patient burned <u>his</u> arm.
The child hit <u>her</u> head.

However, in medical English it is sometimes convenient to objectify these nouns, and this is done with the article *the:*

In this patient, <u>the</u> heart is still quite strong.
<u>The</u> mouth is covered with sores.

Adverbial *the*.   Adverbial *the* is used before comparatives (usually with *all*) to indicate degree or amount:

The sun's brightness makes it all <u>the more difficult</u> to study with an
   optical telescope.
Radiation in the damaged reactor vessel rendered it <u>the harder</u> to
   repair.

Adverbial *the* is also used to indicate a direct or an inverse relationship (see also "Comparisons," Section 3 of this unit):

Direct:
<u>The</u> faster a car moves, <u>the</u> more time it takes to stop.
<u>The</u> higher <u>the</u> birth rate, <u>the</u> greater <u>the</u> population.

Indirect:
The more oil we use, the less there will be for future generations.
The brighter the light, the smaller the lens aperture.

Such a comparison is one of the few sentences in English that can be written without a main verb.

EXERCISE 1.3   Add *the* to the following sentences **where necessary**.

Example: This muscle is called biceps.
  Answer: This muscle is called the biceps.

1. Hepatitis is a disease of liver.
2. The fog made it all more difficult for the plane to land.
3. An optometrist will examine eyes before prescribing corrective lenses.
4. Higher temperature, greater speed of sound through air.
5. Mumps is a childhood illness caused by a virus and characterized by fever and swelling of parotid glands.
6. The surgeons redirected the blood flow before they could open heart.
7. Less people smoke, better are their chances of survival.
8. Flu killed many Amazonian Indians, who had no resistance to the disease.
9. The high quality of the crude oil made the distillation all easier.
10. Boyle's law states that at a constant temperature, smaller volume, greater pressure.

Idiomatic Structures with Ø

The zero article is used in place of *the* when a countable noun is not specialized. This occurs with (1) time expressions and events, (2) institutions, (3) means of transportation, and (4) unfocused singular countable generic nouns.

Time expressions and events.   The zero article occurs in prepositional phrases with day and night, seasons, and meals. However, seasons may also be written with *the*.

day/night	Most people work by <u>day</u> and sleep at <u>night</u>.
seasons	It rains in California in (the) <u>winter</u>.
meals	The patient fainted after <u>dinner</u>.

Institutions (instead of buildings.   See also Unit III, Section 3, "The Prepositions at, on, and in: At").

school	She went to <u>school</u> at the Sorbonne.
work	There was not much progress at <u>work</u>.
church	He goes to <u>church</u> once a week.
home	He works at <u>home</u>.

Means of transportation. (See also Unit II, Section 4, "*How*-Agents.") The team traveled

by car.	by air.	by train.
by bus.	by sea.	by boat.
by plane.	by land.	by donkey.

Unfocused singular countable generic nouns.

Food is classified as <u>carbohydrate</u>, fat, and <u>protein</u>.
Ethyl acetate smells like <u>banana</u>.
There's a movie on <u>television</u>. (See Unit II, Section 1, "Second Mention Without First Mention: Cultural Shared Knowledge.")

The zero article also replaces *the* (1) with a rank, title, or unique post; (2) after noun phrases with *be* or a "naming" verb; and (3) with certain traditional phrases.

Rank, title, or unique post.

Dr. Ernest Lawrence, <u>founder</u> of the Lawrence Berkeley Lab, invented the cyclotron.
Mrs. Pirelli, <u>director</u> of the institute, welcomed the visitors.
Professor Waters, <u>chairman</u> of the civil engineering department, presented his latest research.

Noun phrases after *be* or "naming" verbs (*appoint, declare, elect*).

Dr. Packer <u>is</u> <u>chairman</u> of the physiology department.
The president <u>appointed</u> her <u>director</u> of the Office of Science and Technology.

Certain traditional phrases. Examples:

in fact	The center of the earth's core is, <u>in fact</u>, solid. (See Unit IV, Section 2, "The AND Group.")
in case	Break the glass <u>in case</u> of fire.
at last	The programmers worked until <u>at last</u> they found the bug in the program.
day by day	The patient is improving <u>day by day</u>.

EXERCISE 1.4 Add the correct article plus the noun in parentheses to the blanks.

Example: Break the glass only in (case) _____ of fire.
 Answer: Break the glass only in case of fire.

1. The greatest snowfall occurred in (spring) _____ of 1921.
2. E. O. Lawrence was named (director) _____ of the U.C. Radiation Laboratory in 1936.
3. The bobcat sleeps by (day) _____ and hunts at (night) _____ .
4. Many cold-blooded animals hibernate during (winter) _____ .
5. All first-class mail in the U.S. is delivered by (plane) _____ .
6. In 1914, Einstein was made (professor) _____ at the Prussian Academy of Science in Berlin.
7. The boy, who was hit by (car) _____ , was rushed to the emergency room.
8. This antibiotic should be taken with (lunch) _____ or (dinner) _____ .
9. Students should not go to (school) _____ , nor people to (work) _____ , if they have a cold; rather, they should stay at (home) _____ .
10. Scientists are generally praised for the knowledge they bring to the world. However, in (case) _____ of Galileo, the opposite was true; in 1632, he was forced to renounce his ideas.

Section 2
SENTENCE COMBINING

INFINITIVE STRUCTURES AND *that*-COMPLEMENTS

A special group of verbs that occurs frequently in scientific English allows two different kinds of sentence combining: infinitive structures and *that*-complements. We studied sentence combining with infinitive structures in Unit III, Section 4. A *that*-complement is a sentence that has been made into a noun clause by the addition of the word *that* (see Unit II, Section 5, "Noun-Phrase Parallelism: Types of Noun Phrases"). Look at this example:

Scientists know X.
+X = The planet Neptune has rings.

Scientists know that the planet Neptune has rings.

The word *that* can be deleted only when the sentence is not too complex:

> Scientists know the planet Neptune has rings.
> Scientists know that stars produce their energy by the thermonuclear fusion of hydrogen into helium atoms.

The limited group of verbs with which we are concerned includes verbs used in experimentation and observation, such as *assume, believe, demonstrate, determine, expect, find, report, reveal,* and *show.* These verbs allow either an infinitive structure or a *that*-complement at the end of the sentence. Look at these examples:

> Ptolemy believed X.
>
> $+$ X $=$ $\overbrace{\text{The earth}}^{\text{OBJ.}}$ $\overbrace{\text{was}}^{\text{V}_{\text{inf}} \text{ (to be)}}$ the center of the solar system.
>
> (1)Ptolemy believed the earth to be the center of the solar system.
>
> Ptolemy believed X.
> $+$ X $=$ The earth was the center of the solar system.
>
> (2) Ptolemy believed that the earth was the center of the solar system.

Notice in sentence 1 that the infinitive verb is the result of a special grammatical operation in which the subject of the X sentence *(the earth)* becomes the object of the main sentence *(Ptolemy believed the earth . . . .).* When the object is moved in this way, the verb no longer has a subject and therefore becomes infinitive *(to be).* Here is another example:

> Doctors expect X.
> $+$ X $=$ The heart patient will live for several years.
>
> (1) Doctors expect the heart patient to live for several years.
> (2) Doctors expect that the heart patient will live for several years.

It is often preferable in scientific writing **not** to indicate the experimenter or observer. Sentence 1 with the infinitive structure can therefore be rewritten as an agent-less passive sentence (see Unit II, Section 4):

> The heart patient is expected to live for several years.

For the same reason, sentence 2 with the *that*-complement can be rewritten with *It* and a passive verb form:

> It is expected that the heart patient will live for several years.

EXERCISE 2.1   Complete the following sentences using infinitive structures and *that*-complements.

Example: Researchers have found X. (X = Caffeine decreases blood flow to the brain.)
　　　　　　a. Researchers (that) ————————————————.
　　　　　　b. Caffeine ————————————————.
　　　　　　c. It has been found that ————————————————.
Answers:　a. Researchers have found that caffeine decreases blood flow to the brain.
　　　　　　b. Caffeine has been found to decrease blood flow to the brain.
　　　　　　c. It has been found that caffeine decreases blood flow to the brain.

　　　　1. Investigators believe X. (X = The experiment is quite dangerous.)
　　　　　　a. Investigators (to) ————————————————
　　　　　　————————————————————————.

　　　　　　b. The experiment ————————————————
　　　　　　————————————————————————.

　　　　　　c. It is believed that ————————————————
　　　　　　————————————————————————.

　　　　2. Many scientists assumed X. (X = The shroud of Turin was a fake.)
　　　　　　a. Many scientists (that) ————————————————
　　　　　　————————————————————————.

　　　　　　b. The shroud of Turin ————————————————
　　　　　　————————————————————————.

　　　　　　c. It was assumed that ————————————————
　　　　　　————————————————————————.

　　　　3. Bioengineers expect X. (X = Genetic engineering will provide abundant quantities of now rare and expensive materials.)
　　　　　　a. Bioengineers (to) ————————————————
　　　　　　————————————————————————.

　　　　　　b. Genetic engineering ————————————————
　　　　　　————————————————————————.

　　　　　　c. It is expected that ————————————————
　　　　　　————————————————————————.

　　　　4. Herpetologists have found X. (X = Some snakes have hinged teeth.)
　　　　　　a. Herpetologists ————————————————
　　　　　　————————————————————————.

　　　　　　b. Some snakes ————————————————
　　　　　　————————————————————————.

　　　　　　c. It has been found ————————————————
　　　　　　————————————————————————.

　　　　5. Medical researchers have reported X. (X = X rays may cause heart disease.)
　　　　　　a. Medical researchers ————————————————
　　　　　　————————————————————————.

b. _____
_____.

c. It _____
_____.

6. The U.S. petroleum industry has determined X. (X=Geochemistry helps in finding oil and gas.)

a. _____
_____.

b. _____
_____.

c. _____
_____.

7. Researchers have shown X. (X=Cockroaches are as energy-efficient as most vertebrates.)

a. _____
_____.

b. _____
_____.

c. _____
_____.

8. University of Illinois physiology professor/demonstrate/teenage drinking/delay/maturity/in male adolescents

a. _____
_____.

b. _____
_____.

c. _____
_____.

9. Radio telescopes/reveal/the night sky/contain/many more objects than can be seen by visible-light telescopes

a. _____
_____.

b. _____
_____.

c. _____
_____.

10. A federal study/report/human tissue/contain/a wide range of toxic pesticides

a. _____
_____.

b. _____
_____.

c. _____
_____.

Section 3
MODIFICATION AND REDUCTION

COMPARISONS

Comparisons are necessary in a feasibility study because a feasible choice is based on a comparison of alternatives. Since there are many ways to compare in English, we will discuss comparisons in terms of (1) similarity and (2) difference.

Similarity

Similarity ranges from absolute identity to likeness. Phrases that express similarity may be ranked accordingly.

Absolute identity.

Phrase	Example
be identical (to)	The number of isotopes of chlorine is identical to the number of isotopes of copper. [Cl = 11, Cu = 11]
the same [NOUN] as	Chlorine has the same number of isotopes as copper.
as [ADJ/ADV] as	Chlorine is as isotope-rich as copper.
as	As is the case with copper, chlorine has 11 isotopes.
be alike	Chlorine and copper are alike in the number of isotopes they have.

Approximate identity.

Phrase	Example
be almost identical to	The specific gravity of gold is almost identical to the specific gravity of tungsten. [Au = 19.32, W = 19.30]
almost the same as	The specific gravity of gold is almost the same as that of tungsten.
almost as [ADJ/ADV] as	The specific gravity of tungsten is almost as high as that of gold.
be almost alike	Gold and tungsten are almost alike in their specific gravities.
be like	Gold is like tungsten as far as specific gravity is concerned.
a great resemblance between	There is a great resemblance between gold and tungsten in terms of specific gravity.
a great similarity between	There is a great similarity between the specific gravity of gold and that of tungsten.

Likeness.

Phrase	Example
be almost like	The melting point of sulfur <u>is almost like</u> the melting point of iodine. [S = 112.8, I = 113.5]
a resemblance between	There is <u>a resemblance between</u> the melting point of sulfur and that of iodine.
resemble	Sulfur <u>resembles</u> iodine in terms of melting point.
a similarity between	There is <u>a similarity between</u> the melting points of sulfur and iodine.
be similar to	Iodine <u>is similar to</u> sulfur as far as melting point is concerned.
be close to	The melting point of sulfur <u>is close to</u> that of iodine.

EXERCISE 3.1   With the information from the picture below, fill the blanks with appropriate words or phrases from the list that follows. Use all possible words or phrases in each blank.

Tokyo

pollution level
45 ppm

Los Angeles

pollution level
45 ppm

London

pollution level
45 ppm

as	resemblance
(almost) as X as	resemble
be alike	be similar (to)
be identical (to)	(great) similarity
like	the same (as)

Example: London _____ Los Angeles in its level of pollution.
  Answer: London resembles (is similar to) Los Angeles in its level of pollution.

1. The level of pollution in Tokyo is _____ that in Los Angeles.
2. London and Tokyo _____ in their levels of pollution.
3. Tokyo is _____ Los Angeles as far as level of pollution is concerned.
4. _____ in Tokyo, Los Angeles has a pollution level of 45 parts per million.
5. London and Los Angeles have _____ pollution levels.
6. Tokyo and Los Angeles _____ in their levels of pollution.
7. One _____ between Los Angeles and Tokyo is the level of pollution.
8. London is _____ Los Angeles.
9. Between Tokyo and London, there is a _____ in the level of pollution that exists.
10. Tokyo is _____ Los Angeles.

Difference

Difference ranges from absolute dissimilarity to slight dissimilarity. Phrases that express difference can be ranked accordingly. The difference between the rankings is sometimes only a matter of a qualifying word, such as *slight, a little,* or *much* (see also Unit II, Section 5, "Modifying the Basic Dimension Statement").

Absolute dissimilarity.

Phrase	Example
be (completely) unlike	The superficial qualities of diamond <u>are completely unlike</u> those of graphite. [diamond = hard, crystalline, transparent; graphite = soft, black, opaque]
as opposed to	<u>As opposed to</u> diamond, graphite is a soft, black, opaque substance.
in contrast with	<u>In contrast with</u> graphite, diamond is a hard, transparent, crystalline substance.

Considerable dissimilarity.

Phrase	Example
differ greatly from	The relative atomic mass of aluminum <u>differs greatly from</u> that of lead. [Al = 26.98, Pb = 207.19]

a (great) difference between	There is <u>a great difference between</u> the relative atomic mass of aluminum and that of lead.
be very different from	Lead <u>is very different from</u> aluminum in relative atomic mass.
not nearly as [ADJ/ADV] as	Aluminum is <u>not nearly as heavy as</u> lead.
much [ADJ/ADV + er] than	Lead is <u>much heavier than</u> aluminum.
much more [ADJ/ADV] than	Aluminum is <u>much more chemically active than</u> lead.
much less [ADJ/ADV] than	Aluminum is <u>much less dense</u> than lead.
more [NOUN] than	Aluminum has <u>more electrons in the outer shell than</u> lead.
less [UNCOUNTABLE NOUN] than	Aluminum has <u>less weight than</u> lead.
fewer [COUNTABLE NOUN] than	Aluminum has <u>fewer electrons in the outer shell than</u> lead.

Slight dissimilarity.

Phrase	Example
differ from	Hydrogen <u>differs from</u> helium in boiling point. [H = − 252° F, He = − 268° F]
a (slight) difference between	There is <u>a slight difference between</u> the boiling point of hydrogen and that of helium.
be (somewhat) different from	The boiling point of hydrogen <u>is somewhat different from</u> that of helium.
not (quite) as [ADJ/ADV] as	The boiling point of hydrogen is <u>not quite as low as</u> that of helium.
[ADJ/ADV] + er than	The boiling point of helium is <u>lower than</u> that of hydrogen.
less [ADJ/ADV] than	Helium is <u>less reactive than</u> hydrogen.
a few/a little more [NOUN] than	Helium has <u>a few more isotopes than</u> hydrogen.
a little less [UNCOUNTABLE NOUN] than	Helium has <u>a little less buoyancy than</u> hydrogen.
fewer [COUNTABLE NOUN] than	Hydrogen has <u>fewer electrons in the outer shell than</u> helium.

EXERCISE 3.2   With the information from the picture below, fill the blanks with appropriate words or phrases from the list that follows. Use all possible phrases in each blank.

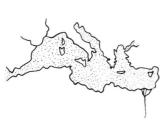
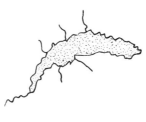

The Mediterranean Sea
NaCl%: normal

Lake Geneva
NaCl%: none

The Dead Sea
NaCl%: high

as opposed to   (great) difference
(completely) unlike   in contrast with
contrasting X with   not (nearly) as X as
differ (greatly) from   be (very) different from
Adjectives: X-er (than); more X (than); less X (than)
Nouns: more + all nouns; less + uncountable nouns; fewer + countable nouns

Example: The Mediterranean Sea is _____ the Dead Sea.
Answer: The Mediterranean Sea is less salty than (not as salty as) the Dead Sea.

1. The Mediterranean Sea _____ the Dead Sea in that the Dead Sea is _____ salty.
2. The Dead Sea is _____ the Mediterranean Sea.
3. The Mediterranean Sea has _____ salt molecules per unit of water _____ the Dead Sea.
4. The Dead Sea _____ the Mediterranean Sea in salt content.
5. The _____ between the Dead Sea and the Mediterranean Sea is that the latter is _____ .
6. The Dead Sea contains _____ salt _____ the Mediterranean Sea.
7. _____ the Mediterranean Sea, Lake Geneva contains _____ salt.
8. The Mediterranean Sea contains _____ salt _____ the Dead Sea.
9. _____ the Dead Sea _____ the Mediterranean Sea, solar-pond engineers have found the former preferable for electric-power generation because of its _____ salt content.
10. Lake Geneva is _____ salty _____ the Dead Sea.

NEGATIVE NOUN COMPARISONS

When making negative noun comparisons, we must use the article *a(n)* with measurement nouns (see Unit II, Section 5, "Dimension Statements and Measurement Nouns"). Look at these examples:

> Alcohol does not have as high <u>a boiling point</u> as water.
> Oxygen does not have as low <u>a relative atomic mass</u> as nitrogen.

The most common adjectives used in negative noun comparisons are *great, high,* and *low. Great* occurs most frequently (not *big*). It is used with all measurement nouns except those that indicate a relative scale. *Great* does not have an antonym in this usage.

> great accuracy
> a great capacity
> great force
> a great population
> a great quantity
> a great span

High is used with measurement nouns that indicate a relative scale—that is, those that indicate degrees of measurement. *High* usually implies a comparison with its antonym *low,* which is used for comparing relative smallness:

high/low boiling point	high/low rate
high/low density	high/low ratio
high/low gravity	high/low specific gravity
high/low growth	high/low temperature
high/low index of refraction	high/low viscosity
high/low level	

Some measurement nouns can be modified with both *great* and *high,* depending on whether or not a degree of measurement is intended. Look at these examples:

> The cortex of the sun is an area of <u>great pressure</u>.
> <u>High pressure</u> will cause the valve to explode (i.e., low pressure will not).
> A <u>great number</u> of researchers are working on a cure for cancer.
> A <u>high number</u> of cancer patients are being cured (i.e., compared with the low number a few years ago).

We do not have to be concerned with the difference between *great* and *high* when the adjectives *many* or *more* are used:

Zinc does not have as <u>many</u> isotopes as silver.
Saturn does not have as <u>much</u> mass as Jupiter.

EXERCISE 3.3   Write out the following negative comparisons using the verb *to have*.

Example: magnesium: atomic number = 12
manganese: atomic number = 25
Answer: Magnesium does not have as high an atomic number as manganese.

1. Mt. Aconcagua (Argentina): height = 7021 meters
   Mt. Dunagiri (India): height = 7066 meters
2. Saudi Arabia: population = 8.35 million
   Mexico: population = 71.9 million
3. carbon (C): melting point = 3550° C
   tungsten (W): melting point = 3410° C
4. the Sea of Japan: average depth = 1350 meters
   the Arctic Ocean: average depth = 1205 meters
5. cirrus clouds: level = 20,000+ feet
   cumulus clouds: level = 6000–20,000 feet
6. lion: speed = 50 mph
   zebra: speed = 40 mph
7. Aswan Dam (Egypt): rated capacity = 2100 Mw
   Kariba Dam (Zimbabwe-Zambia): rated capacity = 1566 Mw
8. water: index of refraction = 1.333
   glass: index of refraction = 1.5
9. kerosene: viscosity at 100° F = 3 X 10^{-4} lbs/ft²
   water: viscosity at 100° F = 1.5 X 10^{-5} lbs/ft²
10. Aldeberan (star): magnitude = 1.1
    Sirius (star): magnitude = − 1.4 (stars with negative numbers are brighter)

NOUN COMPOUNDS

Noun compounds consist of two or more nouns placed together to represent specific items or substances. They represent the ultimate reduction of a relative clause. Look at these examples:

relative clause: The machinist produced <u>a gear that is shaped like a worm.</u>
reduced relative clause: The machinist produced <u>a gear shaped like a worm.</u>
verb → ADJ: The machinist produced <u>a worm-shaped gear.</u>

noun compound:	The machinist produced <u>a worm gear</u>.
relative clause:	Engineers installed <u>a system that is for the purification of water</u>.
reduced relative clause:	Engineers installed <u>a system for the purification of water</u>.
of-phrase inversion:	Engineers installed <u>a system for water purification</u>.
noun compound:	Engineers installed <u>a water-purification system</u>.

Notice that the initial relative clauses in both examples are formal definitions. Noun compounds are actually compressed formal definitions. They can usually be interpreted by reversing the order of the words in the noun compound and inserting prepositions.

1   2   3   3   2   1

a water-purification system = a system for the purification of water

Noun compounds are used frequently in scientific writing because they express a great deal of information with a few words. They are like reminders of a once-learned definition. For this reason, they are difficult for anybody who is not familiar with the original definition, especially when they contain three or more nouns.

EXERCISE 3.4   Choose the correct definition for the noun compound on the left.

Example: worm gear a. a worm that lives in gears
                  (b.) a gear shaped like a worm

    1. test data a. data from a test
                 b. a test of current data
    2. a camera platform a. a platform for a camera
                          b. a camera that sits on a platform
    3. glass fibers a. a kind of glass in the form of fibers
                  b. a kind of fiber made from glass
    4. voltage regulation a. normal voltage as prescribed by regulations
                        b. the regulation of voltage
    5. a radar scan a. a kind of radar that scans
                  b. a scan performed by radar
    6. a carrier lipid a. a lipid that carries peptide units
                   b. a carrier for lipids
    7. a silicon crystal a. silicon in its crystallized form
                    b. a crystal of silicon
    8. surface metal a. metal at the surface
                  b. the surface of a metal

9. process research   a. research into the nature of a process
                        b. the process of doing research
10. a junction diode   a. a junction where two or more diodes come together
                        b. a diode formed from the junction of p- and n-type semiconductors

Formal definitions in English can be classified by the kind of question that the defining relative clause answers about the word being defined (see Unit I, Section 2). These questions are as follows:

1. What are its characteristics?   (properties)
2. What is it composed of?   (material)
3. How does it work?   (operation)
4. What does it do?   (purpose)
5. Where is it used/found?   (location)
6. When is it used?   (time)
7. What does it resemble?   (shape/form)
8. Who discovered/uses it?   (inventor/professional user)

With the exception of the first question, which requires adjectival modification, noun compounds can be classified in the same way:

1. Properties: [requires an adjective structure—e.g., *strong wire*]
2. Material: *copper wire* (wire that is made of copper)
3. Operation: *friction brake* (a brake that works by means of friction)
4. Purpose: *air filter* (a filter for cleaning air)
5. Location: *field mouse* (a mouse that lives in fields)
6. Time: *night hawk* (a hawk that hunts at night)
7. Shape/form: *worm gear* (a gear that is shaped like a worm)
8. Inventor/professional user: *Bunsen burner* (a burner that was invented by Robert Bunsen)

EXERCISE 3.5   Classify the following noun compounds according to the list below. Then define the noun compound in terms of the classification.

[1. properties]   5. location
2. material   6. time
3. operation   7. shape/form
4. purpose   8. inventor/professional user

Example: _____ gear pump
Answer: ___3___ gear pump: a pump that operates by means of gears

1.	_____ long-wire antenna	11.	_____ weed killer
2.	_____ passenger ship	12.	_____ Bernoulli equation
3.	_____ computer industry	13.	_____ morning sickness
4.	_____ furnace gases	14.	_____ ceramic insulator
5.	_____ steam engine	15.	_____ air compressor
6.	_____ water table	16.	_____ needle-nose pliers
7.	_____ annual report	17.	_____ gasoline engine
8.	_____ manganese nodule	18.	_____ vent microbes
9.	_____ vacuum cleaner	19.	_____ day shift
10.	_____ dock crane	20.	_____ blood sample

Longer noun compounds are used in more specialized areas. They can generally be interpreted by reading the compound in reverse.

air-quality program = a program for maintaining the quality of the air
acquired immune deficiency syndrome (AIDS) = a syndrome based on a deficiency of the immune system that is acquired (not inherited)

Noun compounds cannot always be read directly in reverse because certain words have a particularly close relationship. These closely related words are **not** interpreted in reverse order. Look at these examples:

acid-nitrate deposition = the deposition of acid nitrates
coronary heart disease risk = the risk of coronary heart disease
city water chemical contamination monitoring program = a program for monitoring the chemical contamination of city water

In scientific writing, we must be very careful to use only those noun compounds that are acceptable in scientific language. Problems arise if we try to make up new ones (the same is true of abbreviations), especially if they contain three or more nouns.

EXERCISE 3.6   Change the words in italics into noun compounds.

Example: The _vent for air_ should be open.
 Answer: The air vent should be open.

1. _Oil from the peel of citrus fruits_ acts as a pesticide.
2. Neurosurgeons are developing _a map of the system of nerves in humans_.
3. _A soil fumigant made from ethylene dibromide_ has been recently tested.
4. _The risk of lip and throat cancer_ is higher for cigarette smokers.
5. Researchers have located _the site for the binding of RNA_.

6. *The material for insulation that is made from formaldehyde* burned rapidly, releasing toxic fumes.
7. *The facility for electric power that is made of concrete* was seriously damaged by the earthquake.
8. The company announced that it would end *its program of research into superconducting computers.*
9. *The radiation from microwaves that forms a background* resulting from the "big bang" has been determined to be 3° Kelvin.
10. Some stars are fueled by *a cycle of thermonuclear burning involving carbon, nitrogen, and oxygen.*

EXERCISE 3.7   Change the phrases in italics into noun compounds.

Example: The *vent for air* should be open.
 Answer: The air vent should be open.

      A *reduction in noise* of approximately 6 dB could be effected by replacing the existing *assembly containing a blower fan* with *a blower, Model TL-1,* manufactured by Quietaire Corporation of Detroit, and by lining the ducts with Agrafoam, a new *product that performs soundproofing* developed by *the industry that makes automobiles* in West Germany. A further reduction of 1.5 dB could be achieved by replacing *the tiles on the floor made of vinyl* with indoor/outdoor carpet, a practice that has been successful in *centers for the control of traffic in the air.*[1]

EXERCISE 3.8   Make the phrases in italics into noun compounds.

Example: The *vent for air* should be open.
 Answer: The air vent should be open.

      The purpose of the project was to assess the probable *direction of flow at the subsurface* of effluent introduced into the ground. Our scope of work included *explorations in the field, tests of classifications from laboratories,* and a review of *the stratigraphy of the subsurface.* The *explorations in the field* were accomplished by the drilling of four borings using *an auger with a hollow stem that was mounted on a truck.* This report has been prepared for specific application to *the site of the project* in accordance with generally accepted *practices of geotechnical engineering.*[2]

[1]Ron S. Blicq, *Technically–Write!* 2nd. ed., © 1981, p. 146. Reprinted by permission of Prentice-Hall, Inc., Englewood Cliffs, N.J.

[2]Theodore A. Sherman, Simon S. Johnson, *Modern Technical Writing,* 4th ed., © 1983, p. 207. Reprinted by permission of Prentice-Hall, Inc., Englewood Cliffs, N.J.

268   The Research Report and the Feasibility Study

Section 4
VERBS

MODALS

The modals in English include the auxiliary verbs *can, could, may, might, must, shall, should, will, and would* and their various paraphrases. Scientific writing uses these words in more restricted ways than general English. They are most frequently used to indicate (1) obligation, (2) probability, and (3) capability.

Obligation

Since scientific writing is less concerned with moral questions, the sense of obligation is more concerned with the correct way of doing something. The modals that express obligation *(may, should, must, and shall)* may be ranked according to the degree of choice that the human subject (implied or stated) has. (The modals in capital letters are those that occur most frequently in scientific English.)

(Maximum choice) MAY   The compressor system <u>may</u> be insulated.
    ↑   should   The compressor system <u>should</u> be insulated.
    ↓   must   The compressor system <u>must</u> be insulated.
(Minimum choice) shall   The compressor system <u>shall</u> be insulated.

Textbooks frequently use *may* with a passive verb to describe legitimate operations. (The "maximum choice" implied by *may* is also the most polite form.)

    The amount of cholesterol <u>may be determined</u> from a blood sample.
    The pressure <u>may be calculated</u> by means of the formula p = T/V.

The modal *shall* is used by convention in building codes and design specifications. It indicates that the instructions must be followed exactly.

    The condensers <u>shall</u> be adequately sized to operate satisfactorily from 50° F to 110° F ambient temperature.
    The critical speed of the fan wheel <u>shall not</u> be less than 25 percent above the maximum specified speed.

The various grammatical forms of the modals of obligation are shown in the table below. The star (*) indicates that the action described by the verb **did not** actually happen or, with a negative verb, that it **did** actually happen. *Note:* The modal paraphrase of *may* shows the correct sense of *may* but is rarely used (see Answer Key for Unit VI, Exercise 4.2).

MODAL	NEGATIVE	PAST MODAL	PAST NEGATIVE	MODAL PARAPHRASE
may	- - - - - - - - - -	may have	- - - - - - - - - - - - -	(has the option)
should	should not	should have*	should not have*	is recommended
must	must not	had to	did not have to	is required
shall	shall not	had to	did not have to	is required

EXERCISE 4.1   Add the appropriate modal of obligation *(may, should, must, shall)* to the blanks.

Example: A radiation badge _____ be worn in the reactor control room.
  Answer: A radiation badge must be worn in the reactor control room.

1. Researchers concluded that gas stoves _____ be used without adequate ventilation.
2. Cold-box temperatures in the specified refrigeration unit _____ be between $-40°$ and $-80°$ F.
3. The index of refraction for gases _____ be written as

$$n = 1 + \beta \frac{\rho}{\rho s}$$

4. For an exact description of the cost calculations, the reader _____ consult the appendix.
5. Metallic sodium _____ be allowed to come into contact with water; otherwise, it will react violently.
6. All specified ductwork exposed to weather _____ be insulated with three-inch-thick insulation and _____ be weatherproofed.
7. The contrast at any point on the screen _____ be defined as the ratio of the fractional change in illumination to the general illumination.
8. Patients with acute renal failure _____ be given both antibiotics and intravenous and urinary catheters, since dialysis alone will not reduce this complication.
9. It _____ be noted that the deflection of the light ray is a measure of the average density gradient integrated over the X coordinate.
10. Vertical bracing _____ be so arranged that the entire width of all walkway areas has a minimum clear vertical opening of seven feet.

Subjunctive verb form. Modal paraphrases often require the use of the subjunctive base form of the verb (see also Unit III, Section 4, "Infinitive Structures: Infinitives Without *to*") in a *that*-clause (see Section 2 of this unit).

(1) Verb: The hospital REQUIRES that the blood <u>be</u> filtered.
(2) V_{ed2}: It IS REQUIRED that the blood <u>be</u> filtered.
(3) Adj: It is ESSENTIAL that the blood <u>be</u> filtered.
(4) Noun: There is A REQUIREMENT that the blood <u>be</u> filtered.

This verb form is used only with verbs, adjectives, and nouns that **imply** a modal. Sentence 1 above is sometimes informally written:

The hospital requires that the blood <u>must be</u> filtered.

But since the verb *require* implies *must, must* is deleted in formal writing, leaving the base form of the verb:

The hospital requires that the blood <u>be</u> filtered.

Other verbs, adjectives, and nouns that imply modals are listed below.

Modal Implied	Verbs	Adjectives	Nouns
must	ask	compulsory	demand
	command	crucial	direction
	demand	essential	order
	direct	imperative	requirement
	insist	necessary	
	order	obligatory	
	require	vital	
should	desire	advisable	desire
	propose	desirable	proposal
	recommend	fitting	recommendation
	request	preferable	suggestion
	suggest	urgent	
	urge		
may	authorize	permissible	authorization
	permit		

EXERCISE 4.2   Choose five of the sentences in Exercise 4.1 and restate them using modal paraphrases.

Example: A radiation badge must be worn in the reactor control room.
 Answer: It is required that a radiation badge be worn in the reactor control room.

Probability

Probability expresses the degree of certainty that something is correct. The modals that show probability are *must, should, may,* and *might* or *could,* ranging from relative certainty to relative uncertainty. They frequently occur in conclusions and abstracts where the implications of results are discussed.

(Relative	must	The pathogen <u>must</u> be a virus.
certainty)	should	The pathogen <u>should</u> be a virus.
(Relative	may	The pathogen <u>may</u> be a virus.
uncertainty)	might/could	The pathogen <u>might/could</u> be a virus.

The various grammatical forms of the modals of probability are shown in the table below (* = contrary to fact).

MODAL	NEGATIVE	PAST MODAL	PAST NEGATIVE	MODAL PARAPHRASE
must	must	must have	must not have	is certain
should	should not	should have*	should not have*	is likely
may	may not	may have	may not have	is/will perhaps
might/could	might not/ could not	might have/ could have	might not have	is/will possibly

EXERCISE 4.3   Add the appropriate modal of probability *(must, should, may, might/could)* to the blanks.

Example: If the litmus paper turns pink, the solution _____ be acidic.
Answer: If the litmus paper turns pink, the solution must be acidic.

1. The patient _____ died if he had reached the hospital in time.
2. Some studies suggest that certain kinds of habit learning _____ involve the brain's memory system at all.
3. New methods of breaking the bonds of hydrocarbons such as petroleum _____ allow the manipulation of the chemicals into drugs and similarly complex materials.
4. Fleming realized that the bacteria in the Petri dish _____ killed by the penicillin mold nearby.
5. New data suggest that even a modest nuclear war _____ have devastating effects on the atmosphere and global climate.
6. If the absorption spectrum of a star includes the absorption spectrum of hydrogen, that star _____ contain hydrogen.
7. When scattered cumulus clouds are present between 3000 and 10,000 feet above the earth, the ensuing weather _____ be fair.

8. New research suggests that asbestos ＿＿＿＿＿＿＿＿＿＿ be mutagenic itself, but rather enhances other carcinogens.
9. The Titanic ＿＿＿＿＿＿＿＿＿＿ sunk when it hit the iceberg because it was constructed of separate watertight compartments.
10. The rare pygmy chimpanzee ＿＿＿＿＿＿＿＿＿＿ be the best living model of the last common ancestor for apes and humans.

EXERCISE 4.4   Choose five of the sentences from Exercise 4.3 and rewrite them using modal paraphrases.

Example: If the litmus paper turns pink, the solution must be acidic.
 Answer: If the litmus paper turns pink, the solution is certain to be acidic.

Capability

Capability expresses the degree of ability that something or somebody has. The modals that show capability are *can, could* (for hypothetical ability), *should be able to, may be able to,* and *might be able to,* ranging from strong capability to weak capability.

(Strong capability)	CAN	The bridge <u>can</u> support eight tons.
	could	The bridge <u>could</u> support eight tons.
	should be able to	The bridge <u>should be able to</u> support eight tons.
	may be able to	The bridge <u>may be able to</u> support eight tons.
(Weak capability)	might be able to	The bridge <u>might be able to</u> support eight tons.

The various grammatical forms of the modals of capability are shown in the table below (* = contrary to fact).

MODAL	NEGATIVE	PAST MODAL
CAN	cannot	could
could	could not	could have*
should be able to	------------------	should have been able to*
may be able to	may not be able to	may have been able to
might be able to	might not be able to	might have been able to

PAST NEGATIVE	MODAL PARAPHRASE + V$_{ing}$
could not	be capable of
could not have*	would be capable of
--------------------------	should be capable of
may not have been able to	may be capable of
might not have been able to	might be capable of

EXERCISE 4.5   Add the appropriate modal of capability *(can, could, should be able to, may/might be able to)* to the blanks.

Example: A cheetah ———————————— reach a speed of seventy mph.
  Answer: A cheetah can reach a speed of seventy mph.

1. If the cataract operation is a success, the patient ——————————
    see normally in a few days.
2. Spontaneous   electrical   firing ———————————— occur
    anywhere in the heart under certain conditions.
3. If no sterilizer were available, medical instruments ——————————
    be sterilized by chemical means.
4. Since the complete dehydration of 95 percent alcohol requires the
    use of azeotropes and metallic magnesium, it ——————————
    be further concentrated by distillation.
5. Some astronomical physicists have stated that the universe ———————
    expand indefinitely.
6. Hyperkalemia ———————————— occur because of an in-
    crease in hydrogen ions.
7. Astronauts ———————————— control the space shuttle
    manually if the computer navigation system failed.
8. Under the right conditions, the moon ——————————————
    influence earthquakes.
9. This system clearly shows that hormones ——————————————
    exert their physiological effects by altering the specificity of
    enzymes.
10. We ———————————— halt the greenhouse effect if we
    continue to burn fossil fuels at the present rate.

EXERCISE 4.6   Choose five of the sentences in Exercise 4.5 and rewrite
them using modal paraphrases.

Example: A cheetah can reach a speed of seventy mph.
  Answer: A cheetah is able to reach a speed of seventy mph.

Section 5
WRITING AIDS

HEDGING

When reporting the results of their research, scientific writers must be careful
to indicate whether their results are proven facts or probable facts. They do
this by means of hedging, the qualification of the truth of a statement. Hedging is accomplished by means of (1) modals or (2) a statement of probability
with a subordinate clause.

Modals

The modals used in hedging are those concerned with probability (see Section 4 of this unit). Look at this example:

FACT:	Truth Probability
Cancer <u>is</u> caused by a faulty gene.	98–100 percent
HEDGE:	
1. Cancer <u>must be</u> caused by a faulty gene.	80–98 percent
2. Cancer <u>should be</u> caused by a faulty gene.	40–70 percent
3. Cancer <u>may be caused</u> by a faulty gene.	20–40 percent
4. Cancer <u>might/could be</u> caused by a faulty gene.	5–20 percent

The modals *must, should, may,* and *might/could* indicate the decreasing certainty of the statement. Modal paraphrases express the same idea:

FACT:	Truth Probability
Cancer <u>is</u> caused by a virus.	98–100 percent
HEDGE:	
1. Cancer <u>is certain to be</u> caused by a virus.	80–98 percent
2. Cancer <u>is likely to be</u> caused by a virus.	40–70 percent
3. Cancer <u>is perhaps</u> caused by a virus.	20–40 percent
4. Cancer <u>is possibly</u> caused by a virus.	5–20 percent

EXERCISE 5.1   Rewrite the following "facts" using a modal based on the degree of probability indicated.

Example: The center of the Milky Way galaxy is a black hole. (20–40 percent)
Answer: The center of the Milky Way galaxy may be a black hole.

1. The common cold is caused by a virus. (80–98 percent)
2. Anomalons, atomic nuclei that interact with other nuclei more readily than they should, exist. (40–70 percent)
3. Alcoholics use a unique physiological pathway for the breakdown of alcohol. (20–40 percent)
4. Dreaming results from the brain's random firing to debug its overloaded cortex. (5–20 percent)
5. Photosynthesis originated with life itself. (20–40 percent)
6. Fragments of the earth's early crust still exist. (40–70 percent)
7. There is a special genetic marker for brain genes. (80–98 percent)
8. The Mauna Loa volcano on Hawaii will erupt soon. (40–70 percent)

9. Factor VIII, the clotting factor missing from the blood of most hemophiliacs, will be produced by genetic engineering. (5–20 percent)
10. Noise causes cardiovascular-related deaths. (20–40 percent)

EXERCISE 5.2 Rewrite five of the hedging sentences in Exercise 5.3 using modal paraphrases.

Example: The center of the Milky Way galaxy is a black hole. (20–40 percent)
Answer: The center of the Milky Way galaxy is perhaps a black hole.

Statement of Probability with a Subordinate Clause

One way to diminish the boldness of an assertion is to subordinate it (see Unit IV, Section 2, "Coordination and Subordination"). This can be done with (1) a *that*-clause or (2) a passive-infinitive structure (see also "Infinitive Structures and *that*-Complements," Section 2 of this unit).

That-clauses. *That*-clauses that show hedging are commonly attached to main clauses beginning with *there* or *it*.

FACT:
Intelligent life <u>exists</u> elsewhere in the universe.

HEDGE:

There is a {
high probability
good possibility
slight possibility
remote possibility
} that intelligent life exists elsewhere in the universe.

It is {
highly possible
quite possible
possible
remotely possible
} that intelligent life exists elsewhere in the universe.

Passive-infinitive structures. Infinitive structures that show hedging are commonly attached to main clauses with a passive verb of human cognition.

FACT:
Certain people are able to <u>communicate</u> telepathically.
HEDGE:

Certain people are {
known
said
thought
believed
} to be able to communicate telepathically.

EXERCISE 5.3   Rewrite the following "facts" using hedging based on sub-ordination. Be sure to use both *that*-clauses and passive-infinitive structures.

Example: Gasoline fumes cause kidney cancer.
Answers: Gasoline fumes are believed to cause kidney cancer.
It is possible that gasoline fumes cause kidney cancer.
There is a slight possibility that gasoline fumes cause kidney cancer.

1. Mass extinctions occur every 26 million years.
2. The ocean on Neptune's moon Triton is liquid nitrogen.
3. Psychiatrists can predict violent criminal behavior.
4. Fiber-optic lines are the best means of telephone transmission.
5. Certain meteorites contain the chemical bases that combine to form RNA and DNA.
6. Lie detectors are inaccurate.
7. Magnetite particles formed by bacteria cause the sea floor's strong magnetic signals.
8. An asteroid struck the earth 65 million years ago, causing the extinction of the dinosaurs.
9. Acid rain causes serious environmental damage.
10. Low-salt diets reduce high blood pressure.

PART II
Writing a Research Report

Section 6
PREWRITING ACTIVITY: ORGANIZING INFORMATION

A research report is used in scientific writing to present the results of an experiment or study.[3] It is written in such a way that anyone with the proper resources can duplicate the study and verify the findings. The core of the research report is the procedure section, which contains the materials, methods, and results of the experiment. The following exercises will give you practice in preparing the procedure section.

EXERCISE 6.1   Organize the following procedural steps into chronological order.

1. The shear load is increased until the peak strength is reached.
2. The mold is stripped and the specimen is transferred into the shear machine. The upper shear box is set in position and a small normal load is applied. The wires are then cut and the shear load cable is placed in position.
3. The shear stress is determined by dividing the load applied to the specimen by the area of the discontinuity surface.
4. The specimen is now ready for testing and the normal load is increased to the value chosen for the test. The normal load is maintained at a constant level as the shear load is increased.
5. A sample containing the discontinuity to be tested is placed in a mold. The two halves of the rock are wired together to prevent movement, and the sample is then cast in plaster or concrete.
6. This machine is limited to a displacement of approximately one inch. However, since a displacement in excess of this value is required to determine the residual shear strength, the upper half of the specimen must be moved back to its starting position.

[3]*Note to teacher:* For excellent ideas on teaching how to read and write research papers, see Susan S. Hill, Betty F. Soppelsa, and Gregory K. West, "Teaching ESL Students to Read and Write Experimental-Research Papers," *TESOL Quarterly.* September 1982, pp. 333–47.

EXERCISE 6.2   Organize the following procedural steps into chronological order.

1. Soil moisture and leaf-water potential were continuously recorded and adjusted to the original values by rewatering the sand.
2. After four months, the relative longitudinal growth and the survival index (total live plants) were determined.
3. Seedlings of batch X1-3 and batch X1-4 were cultured for two weeks in a growth chamber under the following conditions: 60×10^3 lux, $30°$ C, 12/12 hour photoperiod, model R-500 rooms.
4. Dry weight was determined after drying samples for twenty-four hours at $105°$ C. Water content was calculated from the difference between fresh and dry weight.
5. From each batch, 100 plants were selected and kept under the same environmental conditions.
6. White, washed sand in round 50-by-60-cm pots was used as a substrate.
7. Samples of old and new leaves were detached and used for chemical analysis.

Section 7
STRUCTURE

A typical research report introduces a subject and presents a problem that needs to be addressed, describes a hypothesis for solving the problem, describes an experiment for testing the hypothesis, and describes the implications of the results. Most research reports can thus be divided into four parts: introduction, materials and methods, results, and discussion. The report is always preceded by an abstract and followed by acknowledgments, references, and appendixes (if necessary).

INTRODUCTION

The introduction of a research paper serves as a transition between a general subject area and the particular research of the report. It presents the motivation for the study by describing an existing situation in relation to a problem, need, or inadequacy that must be addressed. It then presents a hypothesis for the resolution of the problem. Look at the following examples:

Example 1

The determination of reliable shear strength parameters is critical in rock slope design. These parameters are susceptible to wide variations due to changes in the characteristics of the jointing system within a rock mass, such as roughness and weathering of joint surfaces. The determination of a representative value for such a parameter requires a procedure that tests a considerable number of samples. A portable shear machine designed for testing small field samples has proven to be simple to operate and to yield satisfactory results. This report describes such an apparatus and the results obtained in a testing program involving 50 field samples of limestone. These results are in agreement with those obtained by Brown and Walton (1979) using a large-scale laboratory shear machine.[4]

Example 2

The desertification of vast extensions of land and the scarcity of water for agricultural purposes have encouraged the development of desert agriculture. Direct watering of plant roots and sophisticated computerized systems have been developed in recent years (Ben Main et al, 1960). Likewise, plastic covers for recovering evaporated water (Ponin, 1964) and similar techniques have been employed. Recently, the hybridization of water-stress-resistant plants with crop plants has become a promising approach (Plin, 1967; Ruscinta, 1973), although success has only been attained with plants with a relatively short growing period. An attempt to interbreed a desert-adapted plant, *Dragonia serpentosa,* a shrub growing in the Muertoese Desert in South America, with *Palatablee aveces,* a forage crop growing in savannahs, has produced a more drought-resistant crop. Neverthless, the properties of this hybrid have not yet been tested. The purpose of this study is to test the nutritional quality and palatability of this hybrid as a possible farm feed.[5]

EXERCISE 7.1   The introduction of a research paper can be described by the formula A but B therefore C, where A is an existing situation, B a problem, and C a hypothesis.[6] Identify these three aspects in the example introductions above.

MATERIALS AND METHODS

The methods-and-materials section describes what materials were used in the experiment and how the data were collected. The methods section is usually presented in chronological order. Examples were given in Exercises 6.1 and 6.2 of this unit.

[4]From a writing assignment by Jose Mendoza entitled "The Use of a Portable Shear Machine in the Determination of Shear Strength Parameters."

[5]From a writing assignment by David Mizrachi entitled "Comparative Drought Resistance of Two Hybrid Forage Cultivars."

[6]This formula was described by Tom Huckin and Leslie Olsen at the 1980 TESOL Convention in San Francisco.

If the precise nature of the materials and methods is especially important, they are sometimes listed with separate subheadings.

Example

MICE

Male and female mice of the following strains were used: BALB/c(d,d); BALB.HTG(d,b); and BALB.B(b,b). These were produced in our own breeding colony from stock obtained from Herman Eisen (Massachusetts Institute of Technology). Letters in parentheses indicate the H.2K and H.2D alleles.

VIRUS

Sendai virus with the inactive fusion glycoprotein (Fo) was grown in Madin-Darby bovine kidney (MDBK) cells, the virus was purified, and Fo was activated by treatment with trypsin (5 g/ml, TPCK-trypsin, Worthington, Freehold, NJ) as described by Scheid and Choppin. Viral infectivity was inactivated by ultraviolet light (UV) as described previously. Stocks of vesicular stomatitis virus were prepared as described previously.

ANTISERA

Anti-Thy-1.2, anti-Ly-1[+] and anti-Ly-2[+] sera were obtained from Cedarlane Laboratories (Hicksville, NY).[7]

RESULTS

The results section shows the data gathered from the experiment, usually as a table or graph with an interpretation. If statistical analysis is required, it is also presented in the results section. However, no calculations should be shown in this section. If it is necessary to include calculations, they are placed in an appendix.

Example

	OD_{15} (mg/l)	OD_5 (mg/l)	$OD_{15} - OD_5$ (mg/l)	BOD (mg/l)
Sample (10 ml)	8.3	5.2	3.1	1.93
Control	8.2	8.1	0.1	—
	OD_{15} = oxygen demand at 15 minutes OD_5 = oxygen demand at 5 days BOD = biochemical oxygen demand			

Figure 1 Oxygen Demand

[7]A. H. Hale, M. J. Roebush, and M. P. McGee, "Cytotoxic T Lymphocytes Specific for Vesicular Stomatitis Virus Recognize the Major Surface Glycoprotein of VSV, 1981," *Antiviral Research*, 1(1981), 64.

The biochemical oxygen demand (Figure 1) indicates the degree of contamination of the park stream. The validity of the results is shown by the fact that (a) OD is greater than 1 mg/l, (b) $OD_{15} - OD_5$ in the control is less than 0.2 mg/l, and (c) $OD_{15} - OD_5$ in the sample is greater than 2 mg/l.[8]

DISCUSSION

The discussion section is concerned with the significance of the results. Some common approaches are to discuss (1) the hypothesis in terms of the results, (2) the experimental design, (3) possible sources of error, (4) the results of other researchers, and/or (5) possible scientific explanations of the results. This is often followed by a descriptiom of the limitations of the study and suggestions for further research.

Example

In general, the possibility of growing crop plants in conditions of reduced water supply is a challenge for further research. Even though the two new hybrids developed are a step in this direction, as the present study shows under laboratory conditions, field research is required to test their real adaptability to natural drought. One troublesome point is the rapid lignification occurring in these hybrids, which reduces their palatability, particularly in hybrid X1-3. Although X1-4 contains less organic matter, it is more suitable for animal food.[9]

ACKNOWLEDGMENTS

In the acknowledgments section, the author thanks the people who helped in the experiment or the writing of the report and gives credit to grants or contracts that made the report possible.

Example 1

Professor George Smith contributed greatly to this paper through his helpful suggestions.

Example 2

This work was supported by a grant from the Bowman Gray School of Medicine, Cancer Center Support (CORE) grant (2-P30-Ca 12197-08), and by grants from NIH, AI 15785, and the American Cancer Society, MV-41.[10]

---

[8]From a writing assignment by Guillermo Ubilla entitled "Biochemical Oxygen Demand."
[9]Mizrachi, "Comparative Drought Resistance."
[10]Hale, Roebush, and McGee, "Cytotoxic T Lymphocytes," p. 69.

REFERENCES

The reference section lists the literature cited and any other sources of information used in the study. This material appears in bibliographical form (see the end of each model in this unit or a style manual for the correct format).

Section 8
MODELS

MODEL 1: AN INVESTIGATION OF THE EFFECT OF RADIOACTIVE LABELING OF DNA ON EXCISION REPAIR IN UV-IRRADIATED HUMAN FIBROBLASTS[11]

Ursula K. Ehmann and Errol C. Friedberg, *Laboratory of Experimental Oncology, Department of Pathology, Stanford University, Stanford, California 94305 U.S.A.*

Abstract

Previous studies on the kinetics of thymine dimer excision and unscheduled DNA synthesis in UV-irradiated human fibroblasts showed a significant discrepancy in these two parameters (Ehmann et al., 1978. *Biophys. J.* **22:** 249). In the present study we have investigated the effect of the level of the radioactive isotope used for labeling cells on the kinetics of a parameter that indirectly measures thymine dimer excision. We find no significant differences in the kinetics of this parameter in cells lightly or heavily labeled with radioactive thymidine.

Introduction

UV-irradiation of living cells at ~254 nm results in the formation of dimers between adjacent intrastrand pyrimidines in DNA (1). We have previously investigated the kinetics of the loss of thymine-containing pyrimidine dimers from the acid-insoluble fraction of several cultured human cell lines and compared these results to the kinetics of repair synthesis measured by autoradiography (2, 3). In those studies we observed that the time of half-maximal loss of dimers ranged from 12 to 22 h after irradiation. In contrast, the time of half-maximal repair synthesis of DNA was ~4.5 h.

We advanced several hypotheses to account for this kinetic discrepancy (2, 3), one of which addressed the possible effects of radioactive labeling of the DNA on cells in culture. Specifically, the measurement of dimer excision by the techniques we employed requires the prelabeling of the DNA of cells with radioactive thymidine for ~24 h before irradiation. On the other hand, the

---

[11]Reproduced from *Biophysical Journal*, 31 (1980), 285–91.

measurement of repair synthesis by autoradiography only utilizes labeling of the DNA after the irradiation, just before sacrifice of the cells. Radioactive labeling of mammalian cells in culture is known to cause DNA strand breaks, mutations, inhibition of cell division, chromosomal aberrations, and cell death (see discussion by Ehmann et al. [2, 4]). We have thus considered the possibility that the use of relatively high levels of radioactive isotope may have a toxic effect on cells in culture that manifests as a reduced efficiency of excision repair. The most direct method for testing this hypothesis would be to compare the kinetics of the loss of thymine-containing pyrimidine dimers in cells exposed to low and high levels of radioactive thymidine. However, in our experience the accurate quantitation of the thymine dimer content of the DNA in cells undergoing active excision (and hence progressively losing dimers) requires a high specific radioactivity of the DNA that can only be achieved by including at least 2μCi/ml and preferably 5–10 μCi/ml of labeled thymidine in the cultures. Such levels of isotope have been shown by others to cause significant perturbations of cellular metabolism in mammalian cells (see reference 2 and references cited therein).

The excision repair of UV-irradiated cells involves both the enzyme-catalyzed incision of DNA at pyrimidine dimer sites and their subsequent excision (5). The efficiency of the former step can be directly measured by quantitating the sensitivity of the cellular DNA to a preparation of enzyme(s) isolated from *Micrococcus luteus* or from phage T4-infected *Escherichia coli* known to catalyze the incision of DNA at pyrimidine dimer sites (6). Sensitivity of the DNA to such enzyme probes indicates that incision of the DNA failed to occur at these sites in the intact cell. The use of this technique (generally referred to as the enzyme-sensitive site [ESS] assay [6]) also requires prelabeling of cellular DNA with a radioisotope; however, a significantly lower specific activity of the DNA can be tolerated than for the direct measurement of the excision of thymine-containing pyrimidine dimers. If prelabeling of the DNA of human cells in culture with radioactive isotope impairs their capacity for excision repair in vivo, one might expect to observe differences in the kinetics of the loss of ESS from DNA of cells exposed to high or low levels of radioactive isotope. The results of such a comparison using normal human fibroblasts are reported here.

Materials and Methods

A normal human diploid fibroblast cell line, GM 316, obtained from the Mutant Cell Repository, Medical Research Institute, Camden, N.J., was cultured in Eagle's Modified Essential Medium as described previously (2). The cells were plated in 100-mm plastic tissue culture petri dishes at a density of 5.7×10^5 cells/dish in 10 ml of growth medium and incubated overnight. The medium was then removed and replaced either with medium containing [³H]thymidine (5.0μCi/ml, 50 Ci/mmol, New England Nuclear, Boston, Mass.) [¹⁴C]thymidine (0.05 μCi/ml, 49 mCi/mmol, New England Nuclear), or medium without label. After 39 h of incubation at 37° C, some of the cells were irradiated with 5 J/m² of 254 nm light at a dose rate of 0.17 J/m²s⁻¹. The cells were then incubated for 0, 3, 6, or 12 h in normal growth medium. At the end of the postirradiation incubation period the cell sheets were washed twice with phosphate buffered saline and incubated for a few minutes with 2 ml of a solution containing 0.05% trypsin plus 0.025% EDTA to remove the cells. To this mixture 7 ml of phosphate buffered saline and 1 ml of growth medium were added to inhibit further trypsin digestion. Clumps of cells were disaggregated by pipetting the cells in the solution 4–5 times with a 10-ml pipette. 9 ml of each sample were used for cell counting and sizing. Cells were counted with a Coulter Model ZBI cell counter, and size spectra for the cell populations were determined with a Coulter Channelizer (Coulter Electronics Inc., Hialeah, Fla.).

For the determination of ESS in cellular DNA, [¹⁴C]thymidine- and [³H]thymidine-labeled cells were mixed and the DNA co-extracted and sedimented on the same gradient. An aliquot of (1.0 ml) of the cell suspension labeled with [³H]thymidine from a particular incubation time was added to 1.0 ml of the suspension labeled with [¹⁴C]thymidine from the same incubation time and the mixture was centrifuged for 7 min at 1,000 rpm in a GLC-1 tabletop centrifuge (Sorvall, Newtown, Conn.). 2 ml of a solution of 0.15 M NaCl and 10 mM EDTA was added to each cell pellet and the mixture was vortexed and recentrifuged. To each pellet 100 µl of an enzyme incubation mixture at pH 7.5 consisting of 0.1 M NaCl, 0.02 M Tris-HCl buffer, pH 7.5, 0.01 M EDTA, 0.01 M β-mercaptoethanol, and 1 mg/ml bovine serum albumin was added. After mixing, the cells were permeabilized by two cycles of rapid freezing and thawing (7). For this procedure the suspensions were frozen by immersing the bottom of the centrifuge tubes in a mixture of dry-ice/acetone for 30 s and then thawed by quickly transferring to a water bath at 37° C for 90 s. An aliquot (10 µl) of a crude extract of *M. luteus* containing saturating amounts of UV-DNA incising activity prepared according to the procedure of Carrier and Setlow (8) was added to each tube of permeabilized cells and the tubes were incubated at 37° C for 15 min (7). After incubation, 0.5 ml of 0.15 M NaCl containing 0.01 M EDTA was added and the suspension was lysed at room temperature for 45 min on the top of 5–20% alkaline sucrose gradients as described by Lett et al. (9). The DNA was centrifuged at 18,000 rpm in a Beckman SW27 rotor (Beckman Instruments, Inc., Spince Div., Palo Alto, Calif.) for 292 min at 20° C, after which the gradients were fractionated from the bottom of the polyallomer centrifuge tubes. The number of average molecular weights and the number of breaks per DNA strand were calculated from the sedimentation profiles as previously described (10). The position of ¹⁴C-labeled T2 phage DNA sedimented on a separate gradient was used as a DNA molecular weight calibration marker.

Results

It has been previously shown by one of us and others (4, 11, 12) that the proliferation of mammalian cells in monolayer culture is retarded by growth in high levels of radioactive thymidine, and that this isotope also causes cell cycle arrest in G_2 phase. Both these effects were observed in the present experiments. GM 316 fibroblasts that were prelabeled for 39 h with either 0.05 μCi/ml of [^{14}C]thymidine (low level of isotope) or 5.0 μCi/ml of [^{3}H]thymidine (high level of isotope) were washed and incubated in fresh medium without the addition of isotope for an additional 12 h. At this time the cell number in the culture exposed to the high level of isotope was about half that observed in the control cultures or the cultures prelabeled with low levels of isotope (Table I). We also observed by direct measurement that the cells treated with a high level of isotope had a median volume more than twice that of controls (Table I, Figure 1). This result is expected of cells blocked late in the cell cycle (13–15). The volume spectrum of the cells labeled with a low level of isotope was slightly larger than that of untreated cells (Table I, Figure 1), indicating that even 0.05 μCi/ml of radioactive thymidine may have caused some perturbation of the cycle. However, this effect was marginal and the growth rate of these cells apparently normal.

After irradiation with 5 J/m^2 of UV light, cells were incubated at 37° C for varying periods of time. The number of sites sensitive to hydrolysis of phosphodiester bonds by the *M. luteus* enzyme preparation was determined by incubating permeabilized cells with the enzyme preparation, sedimenting the DNA through alkaline sucrose gradients and calculating the number average molecular weight and hence the number of enzyme-induced breaks per molecule. The number average molecular weight of the DNA from unirradiated cells labeled with ^{14}C or ^{3}H was 2.1×10^8 and 1.8×10^8 daltons, respectively. These values decreased to 1.8×10^7 and 1.9×10^7, respectively, in samples analyzed immediately after irradiation, indicating the presence of ~9 ESS in the DNA of both ^{14}C- and ^{3}H-labeled cells (Fig. 2). The number of ESS in both the ^{14}C- and ^{3}H-labeled DNA decreased progressively as a function of the time of postirradiation incubation. In repeated experiments the rate of loss of such sites from the ^{3}H-labeled DNA (high level radioactivity) was slightly faster than from the ^{14}C-labeled DNA (low level radioactivity); however, the reverse result was never observed, i.e., we have found no evidence that cells containing heavily labeled DNA are inhibited in the rate of loss of ESS relative to cells containing lightly labeled DNA. The results of a typical experiment are shown in Figs. 2 and 3.

TABLE I
EFFECT OF RADIOISOTOPE ON CELL GROWTH AND CELL VOLUME

CONDITIONS	FINAL CELL COUNT PER DISH ($\times 10^{-5}$)	FINAL MEDIAN CELL VOLUME (RELATIVE UNITS)
No isotope added	12.6	35.4
[^{14}C] (0.05 μCi/ml)	13.6	41.5
[^{3}H] (5.0 μCi/ml)	6.8	78.9

Cells were plated at an initial density of 5.7×10^5/petri dish, incubated for 39 h in the presence or absence of radioactive thymidine and then re-incubated in fresh non-radioactive medium for an additional 12 h before final cell counts and cell volumes were determined. See text for further experimental details.

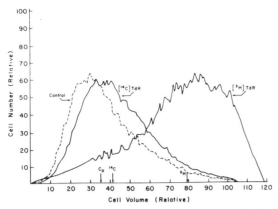

Figure 1 Volume distributions of GM 316 fibroblasts. Cells were grown for 39 h without the addition of radioactive thymidine or in the presence of 0.05 μCi/ml [^{14}C]thymidine or 5.0 μCi/ml [^{3}H]thymidine. The cells were then grown for a further 12 h in medium without isotope and cell volumes were measured in a Coulter Channelizer. The arrows indicate the median values of the cell volume profiles.

Discussion

These experiments demonstrate that the labeling of GM 316 fibroblasts in culture with the high levels of radioactive thymidine used to directly measure excision of thymine-containing pyrimidine dimers has profound effects on cellular metabolism as evidenced by retarded growth rate and by cell cycle arrest. Despite these effects the kinetics of the loss of ESS from the DNA of cells labeled with high levels of radioactivity are not reduced relative to the

Figure 2 Sedimentation profiles of DNA from GM 316 cells centrifuged in alkaline sucrose velocity gradients. See text for experimental details. The arrow marks the position of sedimentation of bacteriophage T2 DNA. Profiles shown with continuous lines are from cells labeled with 5.0 μCi/ml [^{3}H]thymidine (high level radioactivity). Those shown with discontinuous lines represent DNA profiles from cells labeled with 0.05 μCi/ml [^{14}C]thymidine (low level radioactivity).

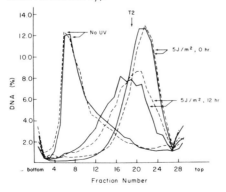

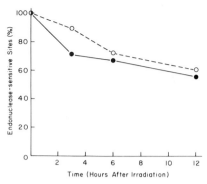

Figure 3 Kinetics of loss of sites in GM 316 DNA sensitive to UV–DNA incising activity from *M. luteus*. See text for experimental details. ●—●, cells labeled with 5.0 μCi/ml [³H]thymidine (high level radioactivity); ○—○, cells labeled with 0.05μCi/ml [¹⁴C]thymidine (low level radioactivity).

kinetics observed with lightly labeled cells. Indeed, the rate of loss of ESS from the former DNA appears to be slightly faster.

There is substantial evidence in the literature (16, 17) indicating that sites in UV-irradiated DNA sensitive to *M. luteus* (enzyme(s) are pyrimidine dimers and that their loss reflects either their incision or incision plus excision during post-UV incubation of cells (18). It is thus evident that the labeling of cells with high levels of radioactive isotope does not inhibit enzymatic events required for incision of DNA in vivo. The use of the ESS assay does not discriminate between sites lost by incision only and those lost by incision followed by excision. Thus we cannot eliminate the possibility that high levels of radioisotope selectively depress the excision of dimers at preincised sites. However, we consider such a selective cellular toxicity unlikely. This consideration is supported by the results of Williams and Cleaver (19) who tested the effect of radiotoxicity on the kinetics of repair synthesis in mammalian cells by pretreating cells with ionizing radiation to mimic chromosomal damage caused by isotope decay. They oberved no significant impairment in the capacity of those cells to carry out subsequent repair synthesis in response to UV radiation damage. Recently, a study on the kinetics of thymine dimer excision and of repair synthesis in human fibroblasts was reported by Konze-Thomas et al. (20). These authors reported no differences in the kinetics of these two parameters. It should be noted however that the experimental protocol (including the level of isotope) and the cell strains used by those authors are different from ours and thus their results are not directly comparable.

In a previous study we suggested four possible explanations for the discrepancy in the kinetics of thymine dimer excision and unscheduled DNA synthesis observed in our hands (2). One hypothesis was that pyrimidine dimers are excised initially as acid-precipitable fragments that are then slowly degraded to acid-soluble products intracellularly. This hypothesis was experimentally tested and negative results obtained (2). A second model proposed was that addressed in the present report. The third hypothesis was that, in human fibroblasts exposed to UV irradiation, incision of DNA adjacent to pyrimidine dimers and repair synthesis precede the actual excision of the dimers, i.e., a "patch and cut" rather than a "cut and patch" mechanism operates. The results of the present experiments directly support this model, since the kinetics of loss of ESS from both ^{3}H- and ^{14}C-labeled DNA (Fig. 3) are faster than the kinetics of the loss of thymine-containing pyrimidine dimers we previously reported (2). A final possibility is that a significant fraction of measured unscheduled DNA synthesis in UV-irradiated cells in culture does not reflect repair synthesis at sites of pyrimidine dimer excision but at other sites in DNA which may or may not be sites of damage.

We thank Dr. R. J. Reynolds for the preparation of extractions of *M. luteus* and for advice in the execution of some of the experiments.

These studies were supported by research grants CA-12428 from the U.S. Public Health Service and NP-174 from the American Cancer Society, as well as by contract EY-76-S-03-0326 with the U.S. Department of Energy. U. K. Ehmann was supported by U.S. Public Health Service Pathology Training grant GM-02236. E. C. Friedberg is the recipient of Research Career Development Award CA-71005 from the U.S. Public Health Service.

Received for publication 6 May 1980.

References

1. SETLOW, J. K. 1966. The molecular basis of biological effects of ultraviolet radiation and photoreactivation. *Curr. Top. Rad. Res.* **2**:193.

2. EHMANN, U. K., K. H. COOK, AND E. C. FRIEDBERG. 1978. The kinetics of thymine dimer excision in ultraviolet-irradiated human cells. *Biophys. J.* **22**:249.

3. EHMANN, U. K., K. H. COOK, AND E. C. FRIEDBERG. 1978. Studies on the molecular mechanisms of nucleotide excision repair in UV-irradiated human cells in culture. *In* DNA Repair Mechanisms. P. C. Hanawalt, E. C. Friedberg, and C. F. Fox, editors. Academic Press, Inc., New York. 315.

4. EHMANN, U. K., J. R. WILLIAMS, W. A. NAGLE, J. A. BROWN, J. A. BELLI, AND J. T. LETT. 1975. Perturbations in cells cycle progression from radioactive DNA precursors. *Nature (Lond.)* **258**:633.

5. FRIEDBERG, E. C., K. H. COOK, J. DUNCAN, AND K. MORTELMANS. 1977. DNA repair enzymes in mammalian cells. *Photochem. Photobiol. Rev.* **2**:263.

6. PATERSON, M. C., P. H. M. LOHMAN, AND M. L. SLUYTER. 1973. Use of a UV-endonuclease from *Micrococcus luteus* to monitor the progress of DNA repair in UV irradiated human cells. *Mutation Res.* **19**:245.

7. VAN ZEELAND, A. A. 1978. Introduction of T4 endonuclease V into frozen and thawed mammalian cells for the determination of removal of UV induced photoproducts. *In* DNA Repair Mechanisms. P. C. Hanawalt, E. C. Friedberg, and C. F. Fox, editors. Academic Press, Inc. N.Y. 307.

8. CARRIER, W. L., AND R. B. SETLOW. 1970. Endonuclease from *Micrococcus luteus* which has activity toward ultraviolet-irradiated deoxyribonucleic acid: purification and properties, *J. Bacteriol.* **102**:178.

9. LETT, J. T., E. S. KLUCIS, AND C. SUN. 1970. On the size of the DNA in the mammalian chromosome: structural subunits. *Biophys. J.* **10**:277.

10. EHMANN, U. K., AND J. T. LETT. 1973. Review and evaluation of molecular weight calculations from the sedimentation profiles of irradiated DNA. *Rad. Res.* **54**:152.

11. POLLACK, A., C. B. BAGWELL, AND G. L. IRVIN, III. 1979. Radiation from tritiated thymidine perturbs the cell cycle progression of stimulated lymphocytes. *Science (Wash. D.C.).***203**:1025.

12. MARZ, R., J. M. ZYLKA, P. G. PLAGEMANN, J. ERBE, R. HOWARD AND J. R. SHEPPARD. 1977. G2+M arrest of cultured mammalian cells after incorporation of tritium-labeled nucleosides. *J. Cell Physiol.* **90**:1.

13. ANDERSON, E. C., G. I. BELL, D. F. PETERSEN, AND R. A. TOBEY, 1969. Cell growth and division. IV. Determination of volume growth rate and division probability. *Biophys. J.* **9**:246.

14. EHMANN, U. K., H. NAGASAWA, D. F. PETERSEN, AND J. T. LETT. 1974. Symptoms of X-ray damage to radiosensitive mouse leukemic cells: asynchronous populations. *Rad. Res.* **60**:453.

15. MEISTRICH, M. L., R. E. MEYN, AND B. BARLOGIE. 1977. Synchronization of mouse L-P59 cells by centrifugal elutriation separation. *Exp. Cell Res.* **105**:169.

16. RIAZUDDIN, S., AND L. GROSSMAN. 1977. *Micrococcus luteus* correndonucleases. I. Resolution and purification of two endonucleases specific for DNA containing pyrimidine dimers. *J. Biol. Chem.* **252**:6280,

17. RIAZUDDIN, S., AND L. GROSSMAN. 1977. *Micrococcus luteus* correndonucleases. II. Mechanism of action of two endonucleases specific for DNA containing pyrimidine dimers. *J. Biol. Chem.* **252**:6287.

18. PATERSON, M. C. 1978. Use of purified lesion-recognizing enzymes to monitor DNA repair *in vivo. Adv. Rad. Biol.* **7**:1–53.

19. WILLIAMS, J. I., AND J. E. CLEAVER. 1978. Excision repair of ultraviolet damage in mammalian cells: evidence for two steps in the excision of pyrimidine dimers. *Biophys. J.* **22**:265.

20. KONZE-THOMAS, B., J. W. LEVINSON, V. M. MAHER, AND J. J. McCORMICK. 1979. Correlation among the rates of dimer excision DNA repair replication and recovery of human cells from potentially lethal damage induced by ultraviolet radiation. *Biophys. J.***28**:315.

Section 9
ANALYSIS

The basic structure of the research report, discussed in detail in Section 7, is as follows:

> Title
> I. Abstract
> II. Introduction
>   A. Background
>   B. Situation-problem-hypothesis (A but B therefore C)
> III. Materials and Methods
>   A. Materials
>   B. Experimental procedure

IV. Results
    A. Data tables (or graphs)
    B. Interpretation
    C. Statistical analysis (optional)
V. Discussion
    A. Significance of results
    B. Suggestions for further research (optional)
VI. Acknowledgments
VII. References
VIII. Appendix (optional)

EXERCISE 9.1   With another student, preferably one who is in your field, analyze Model 1 in Section 8. Determine the function of each subsection (situation-problem-hypothesis, experimental procedure, results, etc.), and label each function in the margin. Discuss your results with your partner. Do you agree?

Section 10
CHOOSING A TOPIC

Your assignment is to write a research report in your own field using the structure shown in Section 9. You have several options:

1. Write about an actual study you have undertaken.
2. Do a small study, perhaps using the students in your class as resources. For example, you could take a standardized measure of a trait or aptitude in the group.
3. Create an imaginary experiment and results and make them into a model research report.

Of course, you are not expected to write a paper that is suitable for publication (unless you are ready for such a step). The purpose of this exercise is to give you practice in using the standard format of a research paper.

PART III
Writing a Feasibility Study

Section 11
PREWRITING ACTIVITY: MAKING A FINANCIAL DECISION

A feasibility study is used in technical writing to recommend that a particular action be taken as a result of a study of various alternatives. It is concerned with both costs and the capability of the recommended action to satisfy the criteria of the study. The following exercises ask you to make a choice based on financial considerations.

EXERCISE 11.1 Read the following problem carefully, and on the basis of financial considerations only, determine if you would recommend or reject the proposed changes.

The Waimea Tile Company in Hawaii has traditionally imported China clay from Europe and the U.S. Now, an engineer has found a method of extracting kaolinite, the main component of China clay, from locally available lava rock. The Waimea Tile Company wants to know if it should invest in expensive new equipment to enable it to use the local lava, or if it should continue to import China clay.

Given data.

Interest rate on borrowed money: 15% per annum
Cost of China clay: $1.00 per pound
Cost of lava rock: $0.04 per pound
Yield of clay from lava rock: 20%
Production of tiles: 200,000 per year
Other production costs (mixing, cutting, drying, packaging): $0.25 per tile
Additional lava-rock production costs (processing lava, etc.): $0.10 per tile
Cost of conversion from imported clay to lava rock (engineering costs, new equipment, other adjustments): $130,000

Profitability criterion.

> Any investment with a break-even period of 7 years or less is accept-
> able to the tile company.

Assumptions.

1. One thousand ceramic tiles require 500 pounds of clay.
2. Time required to convert plant: 18 months.
3. Tile production continues as usual during conversion.
4. Money for total investment must be borrowed at the beginning of
   the construction period.
5. Loan and interest charges must be repaid over 7 years in 84 equal
   payments, which include both principal and interest.

$$\text{Payment} = \text{Loan} \left(\frac{I}{1 - \left(\frac{1}{1+I}\right)^n} \right)$$

I = interest rate per month = $0.15/12$
n = number of months = 84
(For readers who do not have a calculator: $\left(\frac{1}{1+I}\right)^n = 0.35222$.)

EXERCISE 6.2   Read the following problem carefully, and on the basis of
financial considerations only, determine if you would recommend or reject the
proposed changes.[12]

The Atlantic Electric Company, which has always used fresh water to cool its
power facility, is considering the installation of a cooling tower. The company
needs to know if it should (1) buy and install a cooling tower to process the
plant's cooling water load of 1000 gallons per minute, buying makeup water
from the local water company; or (2) continue to buy cooling water from the
water company and discharge it into the sewer system after using it once.

Given data.

Investment

Interest rate on borrowed money: 14% per annum
Purchase price of cooling tower: $33,000
Installation cost: 60.6% of purchase cost
Land for site + site preparation: $15,000
Total investment: $68,000

[12]I am indebted to Dr. David Lyon, formerly of the U. C. Berkeley Department of Chemical Engineering, for the initial form of this problem.

Tower Operating Costs
>   Power: $1.20/hr of operation
>   Labor: $0.80/hr of operation
>   Maintenance: 5% of purchase price per annum
>   Makeup-water requirement:
>   Evaporation and drift: 3% of feed
>   Bleed-off to sewer: 2% of feed

Water Costs
> $0.38/100 ft^3 for first 500 ft^3/month
> $0.50/100 ft^3 for next 49,500 ft^3/month
> $0.43/100 ft^3 for all in excess of 50,000 ft^3/month

Sewer Service Charge
> $0.25/100 ft^3 of discharge to sewer

Profitability criterion.   Any investment with a break-even period of 5 years or less is acceptable to the electric company.

Assumptions.

1. One ft^3 = 7.48 gallons.
2. One month = one twelfth of a year.
3. The plant operates 52 weeks/year, 5 days/week, 16 hours/day.
4. Time required for installation and connection: 6 months.
5. Water is purchased as usual during installation period.
6. Money for total investment must be borrowed at the beginning of the construction period.
7. Loan and interest charges must be repaid over 5 years in 60 equal payments, which include both principal and interest:

$$\text{Payment} = \text{Loan} \left(1 - \frac{1}{\left(\frac{1}{1+I}\right)^{n}} \right)$$

I = interest rate per month = 0.14/12
n = number of months = 60
(For readers who do not have a calculator, $\left(\frac{1}{1+I}\right)^{n} = 0.4986$.)

Section 12
STRUCTURE

A typical feasibility study introduces a subject and presents a problem that needs to be addressed, compares the costs and the capabilities of the alternatives, makes a conclusion based on the results of the comparisons, and recommends that one of the alternatives be implemented. Most feasibility studies can thus be divided into four parts: introduction, cost and capability sections, conclusions, and recommendations. The report is preceded by a cover letter and an abstract and followed by acknowledgments, references, and an appendix containing cost calculations and possibly other materials. The feasibility study, like other technical reports, is set up like a filing system to enable the rapid retrieval of specific information if necessary.

INTRODUCTION

The introduction of a feasibility study indicates the purpose or function of the report (purpose), the circumstances causing it to be written (objective), and, if necessary, the breadth of the report (scope), indicating the limitations of the study. This information can be presented in one paragraph or under subheadings.

Example 1: Introduction[13]

The American Metal Climax, Inc. (AMAX) has holdings at Kirwin and is interested in the future regional ecology of the upper Wood River flowage. The purpose of this report is to assess the possible effects of mining developments on the stream.

Example 2: Introduction[14]

Purpose. The purpose of this report is to recommend that either an ac (alternating current) or dc (direct current) adjustable speed drive be purchased and installed on the number 3-1505 lathe for producing steel crankshafts.

Problem. The General Motors contract calls for 19,500 steel crankshafts. Number 3-1505 is the only lathe large enough to produce such crankshafts, but its constant speed of 1,620 revolutions per minute is good only for cutting softer nodular-iron crankshafts.

Scope. The ac and dc drives are compared according to their cost (break-even point, first cost, rate of capital recovery), and capability (efficiency, overload capacity, braking, availability of parts).

[13]Adapted from Theodore S. Sherman and Simon S. Johnson, *Modern Technical Writing,* 4th ed., © 1983, p. 261. Reprinted by permission of Prentice-Hall, Inc., Englewood Cliffs, N.J.
[14]Steven E. Pauley: *Technical Report Writing Today.* Copyright © 1973 by Houghton Mifflin Company. Used with permission.

COST SECTION

The cost section usually consists of a table presenting a clear comparison of the costs of the alternatives described. An important aspect of such a table is the use of a dash to indicate that a particular cost does not apply. The table must be followed by an interpretation of the chart.

Example:

TABLE 1   COST DATA FOR AC AND DC ADJUSTABLE SPEED DRIVES[15]

ITEM	AC DRIVE	DC DRIVE
First Costs		
Speed drive	$76,000	$110,000
Air damper	1,500	—
Tachometer	2,000	—
Installation	23,000	12,000
Total	$102,000	$122,000
Operational Costs		
Material and labor	$522,175	$520,620
Production losses	325	130
Total	$522,500	$520,750
Maintenance Costs		
Preventive	—	$1,950
Repair	20,000	5,000
Total	$20,000	$6,950

The operational costs are based on the production of 6,500 crankshafts per year. Operational and maintenance costs, unlike first costs, are fixed; in other words, the firm would incur operational and maintenance costs yearly but would pay the first costs only once. Since the ac drive's higher operational and maintenance costs would eventually exceed the dc drive's higher first costs, the dc unit has greater profit potential.

CAPABILITY SECTION

The capability section can consist of graphs and tables comparing the qualities of the alternatives in the feasibility study. On the other hand, it might simply list the different qualities or other considerations by subheading.

Example 1:

Figure 1 shows that the efficiency of the ac and dc drives are very similar and should not affect the choice of systems.[15]

[15]Steven E. Pauley: *Technical Report Writing Today.* Copyright © 1973 by Houghton Mifflin Company. Used with permission.

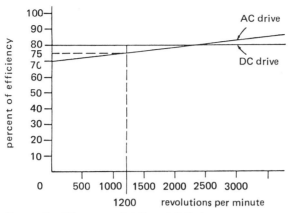

Figure 1 Efficiency of AC and DC drive systems.

Example 2: Comparison of the Zenith X-400 and Monument 65 A Variable-speed Saber Saws[16]

Neither the Zenith X-400 nor the Monument 65 saw requires much effort in feeding. Each has a full-length handle above the motor housing. Each has a slide type of off-on switch that also controls the speed. Each has a rip-and-circle guide adjustable from a minimum of 1 in. to a maximum of 6 in. Each can be adjusted easily to make bevel cuts up to 45 degrees. The differences between them are mainly as follows.

Zenith X-400. Analysis of the Zenith X-400 is based on 25 hours of use of a single example. Other examples were unavailable at the time of the study but the supplier indicated that the model tested was typical.

Superior Features. The Zenith X-400 is relatively quiet and gives less than average trouble with vibration. It handles the sawdust without letting it accumulate on the board and obscure the guide line or blow up into the operator's face. It is provided with a shoe insert to reduce splintering.

Undesirable Features. Although the on-off switch is conveniently located, the device for locking it into position for a selected speed is a little awkward to use. The shape of the shoe does not provide good support when a cut is started or when a thin strip is cut off the edge of the board. It is somewhat heavier than average. Inspection and replacement of the brushes in the motor requires considerable disassembly and reassembly. (Such inspection and possible replacement is important because running the saw with brushes shorter than ¼ in. can damage the motor.)

[16]Theodore S. Sherman and Simon S. Johnson, *Modern Technical Writing*, 4th ed., © 1983, p. 230. Reprinted by permission of Prentice-Hall, Inc., Englewood Cliffs, N.J.

Monument 65 A. Analysis of the Monument 65 A is based on 20 hours of use of five examples. There are neglible differences among the five.

Superior Features. The Monument 65 A is a relatively quiet saw that is easier than average to use. It is light enough that it can be used to cut vertical surfaces without causing more than a minimum of fatigue. It has an auxiliary handle on the left side of the housing. The design of the shoe makes it easier than average for the operator to start a cut and also provides support when cutting a thin strip off the side of the board. It can accommodate an oversize blade that permits cutting lumber up to 6 in. thick.

Undesirable Features. The lightness of the saw results in higher than average vibration. The sawdust has a tendency to collect on the board so that the guide line is sometimes hard to see. It also sometimes blows up into the operator's face. One of the least desirable features is the fact that when the saw must be disassembled, improper positioning of the wire during reassembly causes a danger of electric shock.

CONCLUSION

The conclusion of a feasibility study presents the convictions of the writer based on the evidence presented. It does not, like the recommendation, suggest actions based on those convictions. Some writers prefer to place the conclusion and recommendation directly after the introduction because they believe that the reader wants to see the results before studying the comparisons.

Example: Comparison of AC and DC Adjustable-Speed Drives[17]

1. The initial and installation costs of the dc drive are greater than those of the ac drive. However, the fixed and variable costs associated with the ac drive reduce the latter's total profit-producing capability. The dc drive would have a total profit of $203,000 for the 19,000 crankshafts, which is an increase of 12.8 percent over the present constant-speed drive. The ac drive would have an increase of only 9.8 percent.

2. The capability of the drives is very similar. The main difference between the two is in their braking systems. The dc drive has a regenerative brake which reduces the input power required by the lathe. The braking system of the ac drive is just as quick as the dc, but it is not as efficient, and would cost as much as $3,000 per year more for input power.

RECOMMENDATIONS

The recommendation of a feasibility report suggests actions that should be taken on the basis of the conclusion. It is the reason that the feasibility study exists and is often the only part read by busy managers.

[17]Steven E. Pauley: *Technical Report Writing Today.* Copyright © 1973 Houghton Mifflin Company. Used with permission.

Example: Comparison of AC and DC Adjustable-Speed Drives[18]

1. Because of its profit-making potential, capability, and efficiency, the dc adjustable-speed drive is recommended for installation on the number 3-1505 lathe.
2. The drive should be purchased from the General Electric Company. They have given assurance that the drive can be installed in a maximum of two weeks.
3. A preventative maintenance program should be established by the machine shop. This maintenance program will force continual checking of the drive to see that its components are performing properly.

COVER LETTER

The cover letter is usually attached to the front of the report. It identifies the report, refers to the subject of the report, and states why the report was written. Sometimes it also includes facts that were not appropriate in the report itself.

Example: Memorandum: N.P. Scott Manufacturing Company[19]

To: Mr. Charles Tinkham, Manager
From: Thomas E. McKain, Engineering Department
Date: December 5, 1971
Re: Feasibility Report on Adjustable-Speed Drives

The attached report, entitled "The Feasibility of Adjustable-Speed Drives," is submitted in accordance with your request of October 15, 1971.

The report examines possible adjustable-speed drives to incorporate on the 3-1505 lathe. The cost and capability are compared to those for an alternating current drive.

The more feasible type of drive is recommended. Additional recommendations are made for purchasing the drive and establishing a maintenance program after its installation.

Thomas E. McKain
Thomas E. McKain

ABSTRACT

The writing of abstracts was discussed in detail in Unit V.

[18]Steven E. Pauley: *Technical Report Writing Today.* Copyright © 1973 by Houghton Mifflin Company. Used with permission.
[19]Steven E. Pauley: *Technical Report Writing Today.* Copyright © 1973 by Houghton Mifflin Company. Used with permission.

APPENDIX

The appendix of a feasibility study usually contains the cost calculations of the study. The cost calculations show exactly how you arrived at the figures presented in the cost section of the report. They are included so that readers can check the calculations if they need to.

The appendix might also contain additional material that is useful but not appropriate in the main text of the report, such as detailed test results, lists of equipment, and specifications.

Section 13
MODEL: Selecting New Elevators for the Merrywell Building[20]

H L Winman and Associates

PROFESSIONAL CONSULTING ENGINEERS
475 Reston Avenue-Cleveland, Ohio, 44104

June 17, 19__

Mr. David P. Merrywell, President
Merrywell Enterprises Inc
617 Carswell Avenue
Montrose, OH 45287

Dear Mr. Merrywell:

We enclose our report No. 8-23 "Selecting New Elevators
for the Merrywell Building," which has been prepared in
response to your letter WDR/71/007 dated April 27, 19__.

If you would like us to submit a design for the enlarged
elevator shaft, or to manage the installation project
on your behalf, we shall be glad to be of service.

Sincerely

Barry V. Kingsley

Ian Bailey, P.E.
Head, Civil Engineering

BVK:wp
enc

---

[20]Ron S. Blicq, *Technically—Write! Communicating In a Technological Era*, 2nd ed. © 1981, pp. 192–206. Reprinted by permission of Prentice-Hall, Inc., Englewood Cliffs, N.J.

H L Winman and Associates

PROFESSIONAL CONSULTING ENGINEERS

TECHNICAL REPORT

SELECTING NEW ELEVATORS

FOR

THE MERRYWELL BUILDING

Prepared for:

Mr. David P. Merrywell, President
Merrywell Enterprises Inc
Montrose, Ohio

Prepared by:

Barry V. Kingsley, P.E.
Senior Structural Engineer
H L Winman and Associates

Report No. 8-23

June 17, 19___

SUMMARY

The elevators in the 71-year old Merrywell Building are to be replaced. The new elevators must not only improve the present unsatisfactory elevator service, but must do so within a purchase and installation budget of $500,000.

Of the many types and combinations of elevators considered, the most satisfactory proved to be four 8 ft by 7 ft (2.45 x 2.13 m) deluxe passenger elevators manufactured by the YoYo Elevator Company, one of which will double as a freight elevator during non-peak traffic times. This combination will provide the fast, efficient service requested by the building's tenants for a total price of $450,000, which will be 10% less than the budgeted price.

i

TABLE OF CONTENTS

ii

SELECTING NEW ELEVATORS FOR THE MERRYWELL BUILDING

INTRODUCTION

When in 1954 Merrywell Enterprises Inc purchased the Wescon property in Montrose, Ohio, they renamed it "The Merrywell Building" and renovated the entire exterior and part of the interior. The building's two manually operated passenger elevators and a freight elevator were left intact, although it was recognized that eventually they would have to be replaced.

Recently the elevators have been showing their age. There have been frequent breakdowns and passengers have become increasingly dissatisfied with the in-adequate service provided at peak traffic hours.

In a letter dated April 27, 19__ to H L Winman and Associates, the President of Merrywell Enterprises Inc stated his company's intention to purchase new elevators. He authorized us to evaluate the structural condition of the building, to assess the elevator requirements of the building's occupants, to investigate the types of elevators available, and to recommend the best type or combination of elevators that can be purchased and installed within the proposed budget of $500,000.

CONCLUSIONS

The best combination of elevators that can be installed in the Merrywell Building will be four deluxe 8 ft by 7 ft (2.45 x 2.13 m) passenger models, one of which will serve as a dual-purpose passenger/freight elevator. This selection will provide the fast, efficient service desired by the building's tenants, and will be able to contend with any foreseeable increase in traffic. Its price at $450,000 will be 10% below the proposed budget.

Installation of special elevators requested by some tenants, such as a full-size freight elevator and a small but speedy executive elevator, would be

1

Evaluating Building Condition

We have evaluated the condition of the Merrywell Building and find it to be structurally sound. The underpinning done in 1948 by the previous owner was completely successful and there still are no cracks or signs of further settling. Some additional shoring will be required at the head of the elevator shaft immediately above the 9th floor, but this will be routine work that the elevator manufacturer would expect to do in an old building.

The existing elevator shaft is only 24 feet wide by 8 feet deep (7.35 x 2.45 m), which is unlikely to be large enough for the new elevators. We have therefore investigated relocating the elevators to a different part of the building, or enlarging the existing shaft. Relocation, though possible, would entail major structural alterations and would be very expensive. Enlarging the elevator shaft could be done economically by removing a staircase that runs up the center of the building immediately east of the shaft. This staircase is used very little and its removal would not conflict with fire regulations since there are also fire staircases inside the east and west walls of the building. Removal of the staircase will widen the elevator shaft by 11 feet (3.35 m) which will provide just sufficient space for the new elevators.

Establishing Tenants' Needs

To establish the elevator requirements of the building's tenants we asked a senior executive of each company to answer the questionnaire attached as Appendix A. When we had correlated the answers to all the questionnaires, we identified five significant factors that would have to be considered before selecting the new elevators. (There were also several minor suggestions that we did not include in our analysis, either because they were impractical or because they would have been too costly to incorporate.) The five major factors were:

Every tenant stated that the new elevators must eliminate the lengthy waits that now occur. We carried out a survey at peak travel times and established that passengers waited for their elevators for as much as 70 seconds. Since passengers start becoming impatient after 32 seconds, we estimated that at least three, and probably four, faster passenger elevators would have to be installed to contend with peak-hour traffic.

3

feasible but costly. A freight elevator would restrict passenger-carrying capability, while an executive elevator would elevate the total price to at least 20% above the proposed budget.

The quality and basic prices of elevators built by the major manufacturers are similar. The YoYo Elevator Company has the most attractive quantity price structure and provides the best maintenance service.

The building is structurally sound, although it will require some minor modifications before the new elevators can be installed.

RECOMMENDATIONS

We recommend that four Model C deluxe 8 ft by 7 ft (2.45 x 2.13 m) passenger elevators manufactured by the YoYo Elevator Company be installed in the Merrywell Building. We further recommend that one of these elevators be programmed to provide express passenger service to the top four floors during peak traffic hours, and to serve as a freight elevator at other times.

2

Although all tenants occasionally carry light freight up to their offices, only Rad-Art Graphics and Design Consultants Inc considered that a freight elevator was essential. However, both agreed that a separate freight elevator would not be necessary if one of the new passenger elevators was large enough to carry their displays. They initially quoted 9 feet (2.75 m) as the minimum width they would require, but later conceded that with minor modifications they could reduce the length of their displays to 7 feet 6 inches (2.3 m). All tenants agreed that if a passenger elevator doubled as a freight elevator they would restrict freight movements to non-peak travel times.

The three companies occupying the top four floors of the building requested that one elevator be classified as an express elevator serving only the ground floor and floors 6, 7, 8, and 9. Because these companies represent more than 50% of the building's occupants, we consider that their request is justified.

Three companies expressed a preference for deluxe elevators. Rothesay Mutual Insurance Company, Design Consultants Inc, and Rad-Art Graphics all stated that they had to create an impression of business solidarity in the eyes of their customers, and they felt that deluxe elevators would help to convey this image.

The management of Rothesay Mutual Insurance Company and Vulcan Oil and Fuel Corporation requested that a small key-operated executive elevator be included in our selection for the sole use of top executives of the major companies in the building. We asked other companies to express their views but received only marginal interest. The consensus seemed to be that an executive elevator would have only limited use and the privilege could easily be abused. However, we retained the idea for further evaluation, even though we recognized that an executive elevator would prove costly in terms of passenger usage2.

We decided that the first two of these factors are requirements that must be implemented, while the latter three are preferences that should be incorporated if at all feasible. The controlling influence would be the budget allocation of $500,000 stipulated by the landlord, Merrywell Enterprises Inc. In decreasing order of importance, the requirements are:

1. Passenger waiting time must be no longer than 32 seconds.
2. At least one elevator must be able to accept freight up to 7 ft 6 in. long (2.3 m).
3. An express elevator should serve the top four floors.
4. The elevators should be deluxe models.
5. A small private elevator should be provided for company executives.

Researching Elevator Manufacturers

We asked the three major United States elevator manufacturers to furnish specifications of the elevators they would recommend for the Merrywell Building, plus price quotations and details of their maintenance policies. The catalogs at Appendix B show that except in appearance each company's elevators are basically the same and model for model carry very similar price tags. Significant differences are apparent only in discount policies and maintenance capabilities.

The YoYo Elevator Company of Chicago, Illinois, offers a 10% discount to purchasers contracting for a multiple installation in which all elevators are identical. We queried the other two manufacturers, Jackson Elevators Inc of Detroit, Michigan, and Matson Building Equipment Manufacturers of New York, NY, but neither would agree to incorporate a discount into their contract. Although such a discount agreement might limit the range of elevators from which we could make a selection, we consider that the incentive of a $40,000 to $50,000 reduction in price should not be overlooked.

The YoYo Elevator Company also is the only manufacturer with a service office in Montrose and thus can provide immediate response to calls for emergency maintenance. Both the other manufacturers contract with a local representative to carry out routine maintenance and state that they will fly in a maintenance team within 24 hours if major problems occur. We checked their reputations with the plant superintendents of local buildings using their elevators, and in both cases found that almost invariably delays of three or four days occurred before a maintenance team showed up.

Because of their advantageous pricing policy and superior maintenance service, we consider the new elevators should be selected from the range offered by the YoYo Elevator Company.

Selecting a Suitable Combination of Elevators

There are five basic elevators manufactured by the YoYo Elevator Company which are suitable for installation in the Merrywell Building. These comprise three sizes of passenger elevators that can be supplied in both standard and deluxe versions, a freight elevator, and an executive elevator (see Table 1 for a summary of their specifications, and Appendix C for a more detailed description). From these five basic elevator types we derived numerous combinations of elevators that could be installed in the 35 by 8 ft space (10.7 x 2.45 m), to select a range of elevators which would meet as many as possible of the tenants' requests within the proposed budget of $500,000.

By calculation, we determined that any three of the passenger elevators described in Table 1 could provide an adequate service for the present population of the Merrywell Building, but would be able to handle only a 4%

TABLE 1

Summary of elevators manufactured by the YoYo Elevator Company that are suitable for installation in the Merrywell Building.

MODEL		WIDTH ft (m)	CAPACITY lb (kg)	SPEED (sec/flr)
A	Passenger	6 (1.84)	2000 (910)	5
B	Passenger	7 (2.13)	2600 (1180)	6
C	Passenger	8 (2.45)	3000 (1360)	7
E	Executive Passenger	5 (1.53)	1500 (680)	2
F	Freight	10 (3.05)	5000 (2270)	15

Note: A more detailed description of these elevators is attached as Appendix C.

increase in passenger traffic before again becoming overloaded. Since we estimate that future changes in tenants could possibly increase the population by as much as 23%, we consider that four passenger elevators should be installed. It would be a sounder proposition, both economically and structurally, to anticipate this increase and install the fourth elevator now, than to install it at a later date.

The installation of four passenger elevators would, however, impose a limit on the size of freight elevator that can be installed. Even four of the smallest (model A) passenger elevators would leave only 9 feet (2.75 m) for a freight elevator, thus automatically eliminating the 10-ft wide 5000-lb (3.05 m, 2270 kg) capacity model F freight elevator. The only alternative would be to install one of the largest standard passenger elevators (model C) as a freight elevator, and accept its load restriction of 3000 lb (1360 kg) and maximum interior width of 8 feet (2.45 m). (We have already established that such limitations would be acceptable to the building's tenants.) The price of this combination would, however, exceed the budget by more than $86,000, as shown in Table 2.

To bring the cost down we considered making the model C elevator a dual-purpose unit, serving as a passenger elevator during peak traffic hours, and as a freight elevator at other times. With this arrangement only three regular passenger elevators would have to be purchased, and they could be selected from any of the three models available. Table 2 shows space utilization and the price of the three possible combinations, all of which would cost less than the proposed budget.

TABLE 2

Cost and Space Utilization for Feasible Elevator Combinations

Combination:	1 "C" Freight PLUS 4 "A" Passenger Units	1 "C" Passenger/Freight Unit PLUS Three Passenger Units:		
		3 "A"	3 "B"	3 "C"
Standard Passenger Units	$586,400	$470,400	$480,000	$489,600
Deluxe Passenger Units	$596,800	$478,200	$487,800	$497,400
Total Space Occupied (width)	34.5 ft (10.5 m)	28 ft (8.3 m)	31 ft (8.7 m)	34 ft (10.4 m)

	4 Model C Standard Elevators	4 Model C Deluxe Elevators
List price:	$489,600	$500,000
Discount:	48,960	50,000
Purchase price:	$440,640	$450,000

The combination proposing three model C passenger elevators and one model C passenger/freight elevator fills the available space most efficiently and has the greatest passenger capacity. It also has a significant advantage in that it qualifies for the 10% discount offered by the manufacturer if all elevators installed are exactly the same model and type. By purchasing four identical model C elevators the total purchase price would be:

With this combination the need to provide an express elevator serving the ground floor and floors 6, 7, 8, and 9 is easily satisfied. Probably the best elevator to use would be the "freight" elevator, since it could be programmed to stop only at the preselected floors during peak travel hours, and could revert to its normal role as a freight elevator during off-peak hours (thick protective padding could be used to protect the interior paneling when freight is being carried). A second passenger elevator could be similarly programmed if peak-hour passenger loads dictate more than one express elevator is needed. We do not consider that an express elevator is necessary during non-peak travel times.

The combination of four deluxe model C elevators would therefore satisfy all but one of the tenants' requirements established at the beginning of this report. It reduces waiting time to less than 32 seconds; it provides a

APPENDIX A
to
H L WINMAN AND ASSOCIATES REPORT No. 8-23

QUESTIONNAIRE

To: All Occupants of the Merrywell Building

To assist us in selecting a new range of elevators for the Merrywell Building, please complete the following questionnaire and return it to Barry V Kingsley, H L Winman and Associates, 475 Reston Avenue, Cleveland OH, 44104.

Company Name:..........   Total No. of employees:..........

Floor:..........

1. Do you consider the present elevator service satisfactory? ....
    ....(please elaborate)......................

2. What is the average length of time your employees have to wait for an
    elevator? -- At ground level:....secs. At your floor:.....secs.

3. Do you need a freight elevator? .... (If so, please state dimensions
    and weight of largest loads carried) ....

4. What percentage of your employees' travel is between your floor and
    the ground floor? .....%; between your floor and other floors? .....%

5. Would a small private elevator be of use to your company's executives?
    .... If so, how many members of your staff would use it? ....

6. If you have additional comments or suggestions, please write them here:

                                                Thank you for your cooperation.

---

freight elevator able to handle articles up to 8 ft (2.45 m) long; it provides an express elevator serving the top four floors; and it offers prestige. Not only does it meet the budget appropriation of $500,000 established by Merrywell Enterprises Inc, but it also qualifies for a discount of $50,000 which will bring the total purchase price down to $450,000.

This arrangement does not, however, satisfy the request for a small private elevator for company executives. The only combination in Table 2 that leaves enough room in the elevator shaft to install an executive elevator consists of 3 model A passenger units and 1 model C passenger/freight unit. With the addition of a model E executive elevator, the occupied space would increase to 33.5 feet (10.3 m), and the cost would rise to $613,000 or $620,800, depending on whether standard or deluxe passenger elevators are installed. Although this combination would provide superior service for all of the building's occupants, we consider that its extra cost of at least $113,000 more than the budget, and $163,000 more than the cost of four model C deluxe passenger elevators, is too expensive to warrant the limited extra convenience it offers.

REFERENCES

1. Byron Johnson, "Don't Keep Passengers Waiting" in Elevators
    in the Industrial Complex (New York: Antrim Book Company,
    1971), p. 75.

2. K. K. Krauston, "The Executive's Private Elevator" in
    Business and Industry, 27:5, May 1980, p. 43.

8

APPENDIX C

to

H L WINMAN AND ASSOCIATES REPORT No. 8-23

DESCRIPTIONS OF SUITABLE ELEVATORS

The new elevators will be selected from the following range offered by the YoYo Elevator Manufacturing Company. Standard elevators have brown metal paneling, vinyl-tiled floors, and plain trim on doors and exterior; deluxe units have deep-pile carpeting, mahogany paneling, fancy trim inside and out, and an emergency telephone. All passenger elevators are fully automatic.

TYPE A

Size 6 ft (1.84 m) square; capacity 2000 lb (910 kg); speed 5 seconds per floor. Price: standard unit $116,000; deluxe unit $118,600.

TYPE B

Size 7 ft (2.13 m) square; capacity 2600 lb (1180 kg); speed 6 seconds per floor. Price: standard unit $119,200; deluxe unit $121,800.

TYPE C

Size 8 ft wide x 7 ft deep (2.45 x 2.13 m); capacity 3000 lb (1360 kg); speed 7 seconds per floor. Price: standard unit $122,400; deluxe unit $125,000.

TYPE E (Executive)

Deluxe unit 5 ft (1.53 m) square; capacity 1500 lb (680 kg); very high speed of 2 seconds per floor (on non-stop runs of 6 floors or more). Price: $142,600. Two versions available:

                                  Type EB - push-button operated
                                  Type EK - key operated

TYPE F (Freight)

Rugged unit 10 ft wide x 7 ft 6 in. deep (3.05 x 2.3 m); manually operated; capacity 5000 lb (2270 kg); slow speed (15 seconds per floor). Price: $130,000.

Elevator sizes quoted are exterior dimensions. An additional clearance of 6 in. (15.2 cm) is required between each elevator, and 3 in. (7.6 cm) between each end elevator and the wall. Prices include purchase, installation, and one year's free maintenance.

APPENDIX B

to

H L WINMAN AND ASSOCIATES REPORT No. 8-23

ELEVATOR MANUFACTURERS' CATALOGS

This appendix contains catalogs supplied by the three major North American elevator manufacturers:

1. YoYo Elevator Company, Chicago, Ohio   Page 2

2. Jackson Elevators Inc, Detroit, Michigan   Page 13

3. Matson Building Equipment Manufacturers, New York   Page 27

Price quotations and maintenance policies appear on the final page of each catalog.

NOTE:

Because of their bulk, the catalogs have been omitted from this copy of the report.

1

Section 14
ANALYSIS

The basic structure of a feasibility study, discussed in Section 12, is as follows:

<table>
<tr><td>

I. Preliminary Material
  A. Cover letter
  B. Title page
  C. Abstract
  D. Table of contents
  E. List of illustrations

</td><td>

II. Main Report
  A. Introduction
    1. Purpose
    2. Problem (or objective)
    3. Scope (optional)
  B. Cost section
  C. Capability section
  D. Conclusion
  E. Recommendations
  F. Acknowledgments
  G. References
  H. Appendixes
    1. Cost calculations
    2. Detailed test results, etc.

</td></tr>
</table>

EXERCISE 14.1   With another student, preferably one who is in your field, analyze the model in Section 13. Determine the function of each section in the report (purpose, problem, scope, cost section, capability section, etc.), and label it in the margin. Compare your results with those of your partner.

Section 15
CHOOSING A TOPIC

Your assignment is to write a feasibility study based on the structure discussed in the previous sections. If you already have suitable data for presentation in this form, use it. Otherwise, follow the instructions in Exercise 15.1.

EXERCISE 15.1   Write a feasibility study based on the financial problem you solved in Exercise 11.1 (China clay versus lava rock) or Exercise 11.2 (cooling tower versus public water). The problem will provide the basis of the cost section of your report. You will have to create the capability section yourself, either from research or from common-sense considerations.

APPENDIX A
Irregular Verbs

Verb (Base Form)	Simple Past (Verb$_{ed1}$)	Past Participle (Verb$_{ed2}$)	Verb (Base Form)	Simple Past (Verb$_{ed1}$)	Past Participle (Verb$_{ed2}$)
arise	arose	arisen	lead	led	led
bear	bore	borne (born)	leave	left	left
become	became	become	let	let	let
begin	began	begun	lie	lay	lain
bend	bent	bent	lose	lost	lost
break	broke	broken	make	made	made
bring	brought	brought	mean	meant	meant
build	built	built	put	put	put
catch	caught	caught	read	read	read
choose	chose	chosen	rise	rose	risen
come	came	come	run	ran	run
cut	cut	cut	say	said	said
do	did	done	see	saw	seen
draw	drew	drawn	send	sent	sent
fall	fell	fallen	set	set	set
feed	fed	fed	show	showed	shown
find	found	found	shut	shut	shut
get	got	gotten	spend	spent	spent
give	gave	given	spread	spread	spread
go	went	gone	stand	stood	stood
grow	grew	grown	sweep	swept	swept
have	had	had	take	took	taken
hold	held	held	tear	tore	torn
keep	kept	kept	think	thought	thought
know	knew	known	write	wrote	written

APPENDIX B
Glossary

Key

N	=	NOUN
+	=	COUNTABLE
−	=	UNCOUNTABLE
+ −	=	COUNTABLE OR UNCOUNTABLE
V	=	VERB
I	=	INTRANSITIVE (NO OBJECT)
T	=	TRANSITIVE (OBJECT)
IT	=	INTRANSITIVE AND TRANSITIVE
ADJ	=	ADJECTIVE
ADV	=	ADVERB

abacus   N+   a Chinese counting instrument

abandon   VT   to leave without intending to return

aberration   N+   a deviation from what is normal

abrasion   N+ −   (med) an injury from a rough surface

absorb   VT   to take in

abundant   ADJ   plentiful, more than enough

accelerator   N+   a device for increasing speed

accumulate   VI   to increase in quantity

accuracy   N+ −   degree of freedom from error

acetaldehyde   N− C_2H_4O, derived from the hydration of acetylene

acetic acid   N+   a colorless liquid (CH_3COOH) in vinegar

acid rain   N−   rain containing pollutants

acid-resistant   ADJ   not harmed by acid

acoustic   ADJ   concerning sound or the sense of hearing

acute   ADJ   sharp, severe

additive   N+   a substance added in small amounts

administer   VT   to give or provide in a formal manner

adolescent   N+   a person between childhood and maturity

adrenaline   N−   a hormone that stimulates the nervous system

agent   N+   something that produces a change

ailment   N+   a mild bodily disorder or illness

airtight   ADJ   not allowing air to enter or escape

alarm   N+   an apparatus that gives a warning signal

albumin   N−   a water-soluble protein

alga (pl: -ae)   N−   a water plant with no stems or leaves

algal   ADJ   concerned with algae

aliquot   ADJ   describes numbers that divide without a remainder

alkali   N+ −   a substance that neutralizes acid

alkaloid   N+   a natural chemical used as a drug

alkane   N+   a hydrocarbon containing only single bonds

allele   N+   a gene that can mutate to the form of another gene

alternator   N+   an electric generator that produces alternating current

altitude   N+ −   the height above sea level

Alzheimer's disease   N−   a form of mental infirmity in the aged

amalgam   N+   an alloy of mercury (Hg) and another metal

amber   N−   a hardened, clear yellow-brown resin

ambient   ADJ   surrounding

ambulance   N+   a vehicle for carrying sick or injured people

ammonia   N−   a colorless gas, NH_3

amplifier   N+   a device for increasing voltage

anchorage   N+ −   a place where something can be firmly fixed

aneroid   ADJ   depending on a vacuum

anesthetic   N+   a substance that produces loss of sensation

aneurysm   N+   a localized dilation of an artery

angiosperm   N+   a flowering plant

angstrom   N+ 1×10^{-11} meters

angular velocity   N+ −   the rate of rotation around an axis

annual ring   N+   a circle of yearly tree growth in stem or trunk

anodize   VT   to coat metal with a protective layer

anomalon N+ atomic nuclei that interact with other nuclei too readily

antenna N+ a wire to transmit or receive radio waves

antibiotic N+ a drug that destroys bacteria

antibody N+ a protein formed in the blood in reaction to certain substances that it then destroys

antifreeze N− a substance that prevents freezing

antigen N+ a substance that stimulates the production of an antibody

antiseptic N+ a substance that kills germs

anvil N+ a block on which iron is hammered into shape

aorta N+ the great artery from the left side of the heart

apex N+ the tip, the highest point

apnea N− a sleep disorder

aquo complex N+ a chemical compound in which water is bound

arc N+ − part of a curve

arctic ADJ of the region near the North Pole

argon N− an element (Ar)

armor N− protective covering

arrest VT to stop or prevent the procedure of an action

arsenal N+ a place where weapons are stored or made

arsenic N− an element (As)

artery N+ a tube carrying blood away from the heart

arthritis N− a condition causing pain in the joints

arthropod N+ the phylum containing insects, crustaceans, arachnids, etc.

artificial ADJ not natural

asbestos N− a fibrous mineral used for insulation

aspirin N+ − a common pain-relieving drug

assay N+ a test of quality or composition

asteroid N+ a very small planet circling the sun

astronaut N+ a person trained to travel in a spacecraft

atoll N+ a ring-shaped reef with a lagoon

auger N+ a tool for boring holes in wood

aurora N+ colored light from an electrical effect

aurora borealis N+ the northern lights

autonomic ADJ of that part of the nervous system that functions independently of the will

autonomous ADJ self-regulating

axis N+ the line around which an object rotates

axon N+ a part of a nerve cell

azeotrope N+ a liquid mixture that gives a vapor of the same composition

backbone N+ the vertebral column of bones in the back, the spine

bacteriophage N+ a virus that is a parasite on bacteria

bacterium (pl: -a) N+ a microscopic organism

badge N+ a piece of metal worn on clothing showing rank or status

bait N− food placed to attract prey

balsa N− lightweight wood for making models

bamboo N− giant tropical grass

ban VT to forbid officially

bandage N+ a strip of material for binding a wound

bar chart N+ a graph using columns of varying heights

bark N− the outer covering of a tree

barrel N+ a large cylindrical container

barrier N+ something that prevents or controls progress

batch N+ a number of things prepared at the same time

battery N+ a chemical source of electricity

bauxite N− ore containing aluminum

beaker N+ a glass laboratory container

beam N+ a ray of light; a supporting structure

bearing N+ a device for removing friction where something rotates

bedrock N− solid rock beneath the soil

benzene N− a solvent with the formula C_6H_6

benzine N− a cleaning solvent

beriberi N− a nerve disease caused by a thiamine deficiency

betatron N+ a device for accelerating electrons

bevel cut N+ a cut with a sloping edge

bicameral ADJ having two chambers

biceps N+ the muscle at the front of the upper arm

bicuspid N+ a tooth with two pointed ends (cusps)

bile N− a secretion of the liver that assists in the digestion of fat

bimetallic strip N+ a strip made of two metals that bends when heated

binary ADJ involving two numbers or parts

biosphere N+ the narrow band on earth in which life can exist

biosynthesis N− the production of chemical compounds by living things

black hole N+ a region in space from which no light can escape

blade N+ the cutting part of a knife

blast N+ a sudden strong rush of air or water; an explosion

blastomere N+ one cell formed by the cleavage of an animal zygote

block VT to stop or prevent

blood N− the red oxygen-carrying liquid in animals

blood clot N+ a small, thickened mass of blood

bloom VI to bear flowers

blower N+ a device for moving air

blurred   ADJ   not clear or defined

bobcat   N+   an American lynx (type of wild-cat) with a short tail

boldface   ADJ   describing print or type with heavy dark lines

boll weevil   N+   a small destructive insect in cotton plants

bond   VIT   to connect or join

booster rocket   N+   a rocket that powers a launch and then falls away

bore   N+   the hollow inside of an engine cylinder; a hole made by boring

borer   N+   a device for making a hole in wood

bounce   VIT   to spring back when sent against a hard surface

bovine   ADJ   of, from, or like a cow

bracing   N−   a mechanism of support

bracket   N+   a support projecting from a wall

braid   VT   to interlace three or more strands

brain   N+   the center of the nervous system

brake drum   N+   a cylinder attached to a wheel receiving pressure from the brake shoe

brake line   N+   the metal tube running from the brake pedal to the drum

brake shoe   N+   the part of a brake that comes in contact with the drum

branch   N+   a subdivision; an arm of a tree

break-even point   N+   a point of balance of financial gain and loss

breakdown   N+   an analysis; a failure

breed   VIT   to cause to reproduce

bridge   N+   (dental) a device for connecting a false tooth to real teeth

brine   N−   a salt solution

brittle   ADJ   hard but easily broken

bromine   N−   an element (Br)

brush   N+   electrical conductor in a motor or generator

bubble   N+   a thin ball of liquid enclosing air

buckling   N−   the result of crumpling under pressure

buffer   VT   to lessen the effect of an impact

bulb   N+   the reservoir in a thermometer, usually for mercury

bulldozer   N+   a machine for moving earth

bulletin   N+   a spoken or written announcement

bunch   N+   a number of similar things together

bundle   N+   a collection of things loosely fastened together

Bunsen burner   N+   a device for supplying heat in a lab

buoyancy   N+ −   the ability to float

buret   N+   a graduated glass tube for measuring

burned-out   ADJ   no longer functioning

cable   N+   a thick fiber or wire

caffeine   N−   the stimulant in coffee or tea

calibrated   ADJ   marked with units of measurement

caloric content   N−   the amount of energy food contains

campanulate   ADJ   bell-shaped

camphor   N−   a strong-smelling white substance used in medicine

canal   N+   a channel or passage

canning   N−   the process of preserving food in containers

caoutchouc   N−   a substance found in latex

capillary   N+   the smallest blood vessel; a thin tube

capsule   N+   a small container for medicine or astronauts

carbohydrate   N+   an organic compound containing carbon, hydrogen, and oxygen

carbon   N−   an element (C)

carcass   N+   the dead body of an animal

carcinogen   N+   a cancer-causing substance

carrier   N+   a person or thing that carries

cast   VT   to throw

catalyst   N+   a substance aiding a reaction but not consumed

cataract   N+   a condition in which the eye lens becomes cloudy

catheter   N+   a tube that is inserted into the body

cathode   N+   the electrode by which current leaves a device

cathode ray tube   N+   a vacuum tube that produces an image (CRT)

cattle   N−   large farm animals: cows, bulls, steers

cave   N+   a natural hollow in a cliff or in the ground

cavitation   N−   damage caused by mechanical forces in liquids

cell line   N+   cells with known or controlled heredity

cellulose   N−   an organic substance found in plant tissue

cement   N−   a gray powder that hardens after mixing with water

centrifuge   N+   a machine that separates liquids by spinning

centripetal   ADJ   moving toward the center of an axis

cerebral   ADJ   of the brain

chain reaction   N+   a series of events each of which causes the next

chart   N+   a diagram that gives orderly information

cheetah   N+   a kind of leopard

chew   VT   to work or grind between the teeth

chicken pox   N−   a viral children's disease characterized by red spots on the skin

chimney   N+   an opening for releasing gases

chip   N+   a small piece of silicon with a complex circuit

chlorine   N−   an element (Cl)

chlorofluorocarbon   N+   a chemical occurring in industrial waste

chlorophyll N− the green coloring matter in plants

cholesterol N− a fatty substance in animal tissues

chromium N− an element (Cr)

chromosome N+ the structure that carries genes

chromotography N− determining components of a substance by means of color separation

churn VT to stir or swirl with force

circuit N+ a closed path for electric current

circuitry N− circuits collectively

circumference N+ − the boundary of a circle

citrus fruit N− lemons, oranges, grapefruit, etc.

clay N− sticky earth used for making pots

cliff N+ a steep rock face

clog VT to block

clotting factor N+ the chemical that causes blood to clot

coagulate VIT to become thick

coal gasification N− the process of making coal into gas

cobalt N− an element (Co)

cockroach N+ a common household insect

coefficient N+ a mathematical factor

coil N+ a length of wire wound in a spiral

collagen N− a protein found in bone and tissue

collapse VIT to fall down or in suddenly

collide VI to meet and strike

colon N+ the lower part of the large intestine

comet N+ a fiery object that moves around the sun

compass N+ a device with a needle that points north

compile VT to collect and arrange into a list

compiler N+ a systems program that translates a source program into machine language

compound N+ a substance with more than one part

compression N− forcing something into less space; squeezing

condense VT to make more concentrated

condenser N+ a device for changing gas into liquid

condor N+ a large vulture (bird)

cone N+ a solid body that narrows to a point from a circle

configured (with) ADJ put into a special arrangement

congestive ADJ blocking by accumulation

consciousness N− awareness; the state of being awake

constantan N− an alloy of 60 percent copper and 40 percent nickel

constellation N+ a group of fixed stars; a pattern

constriction N+ a narrow area

consume VT to use up; to eat up; to destroy

consumption N− the act or process of consuming

contaminate VT to pollute

continuum N+ something that extends continuously

contour N+ an outline, a shape

contraceptive N+ a drug or device for preventing pregnancy

contract VT (med) to catch an illness

control subject N+ a standard for checking the results of an experiment

convert VT to change from one form or use to another

cooling tower N+ a structure for removing heat from water

coral N− a hard substance built by tiny sea animals

cord N+ long, thin, twisted strands; an electrical connector

core N+ the central part

core temperature N+ the temperature of the fuel area in a reactor

corium N+ the dermis

cornea N+ the transparent outer covering of the eye

corrosion N− destruction by chemical action

cortex N+ the gray matter of the brain

cosmic ray N+ high-energy radiation from outer space

cotyledon N+ the first leaf growing from a seed

counterclockwise ADJ in a curve from right to left

coyote N+ the North American prairie wolf

crane N+ apparatus for lifting or moving heavy objects

cranium N+ the bones enclosing the brain, the skull

crankshaft N+ the shaft driven by an internal-combustion engine

crater N+ a bowl-shaped cavity

crimp VT to bend and attach

criterion (pl: -a) N+ a standard of judgment

crop N+ a group of plants grown for food

cross N+ a living thing produced by cross-breeding, a hybrid

crude ADJ not refined

crust N+ a hard outer layer

crustal plate N+ one of many flat, thin sheets covering the earth

cryoscope N+ a device for determining the freezing point

crystallize VI to form into a transparent shape

crystallography N− the study of crystal structure

cultivated ADJ prepared and used for crops

culture N+ (bio) a quantity of bacteria grown for study

cumulus N+ a type of cloud formed in a rounded mass

currency N+ − money in use

current N+ water movement; the flow of electricity

curve N+ a line of which no part is straight

cyanide N− a very poisonous substance (usually KCN)

cyclone N+ a violent circular windstorm

cyclosporine N− a drug that reduces the rejection of organ implants

cyclotron N+ a device for accelerating particles

cylinder N+ a chamber in which a piston moves in an engine

cytokinase N+ an enzyme that aids cell division

cytoplasm N− the cellular substance surrounding the nucleus

dairy product N+ a product made from milk

dalton N+ 1.65×10^{-24} grams

deaden VT to anesthetize, to make numb

debug VT to remove errors in a computer program

decibel N+ a unit for measuring the loudness of sound

deciduous ADJ of a tree that sheds its leaves annually

decompose VI to decay, to separate into parts

decongestant N+ a medicine that dries mucous membranes

decontaminate VT to remove radiation or other contaminants

deficiency N+ a lack or shortage

deficit N+ an excess of expenditure over income

deflect VIT to cause to turn aside

deforestation N− the removal of trees

dehydrated ADJ had the moisture removed from

delta N+ the land that forms at the mouth of a river

den N+ the "home" of certain wild animals

dendrite N+ the branched end of a nerve cell

density N+ − the ratio of mass to volume

dentition N+ − the characteristic arrangement of the teeth

denture N+ a set of artificial teeth

deposit VT to leave as a covering of matter

depression N+ − a state of excessive sadness

derivative N+ a thing that is obtained from a source

dermis N+ the layer of tissue below the epidermis

desert N+ dry, usually sand-covered land

desertification N− the process of becoming a desert

deterioration N+ − the process of becoming worse

detonation N+ − the cause of an explosion

deviation N− a turning away from a course of action

devoid (of) ADJ lacking, without

Dewar jar N+ a double-walled glass vessel for cold liquids

diagnosis N+ − a statement of the nature of a disease

dialysis N− the purification of the blood

dicotyledon N+ a flowering plant with two first leaves

die casting N− the process of forming metal in a mold

diet N+ − the food usually eaten

diffract VT to break a beam of light into a spectrum

diffusion N+ the process of spreading throughout

digest VT to dissolve food for absorption

digestible ADJ capable of being digested

digitized ADJ formed into numbers or exact quantities

dike N+ a long wall to prevent flooding

dimer N+ a molecule or compound formed by joining two molecules of the same substance

dinosaur N+ an extinct lizardlike creature

diode N+ a device for making current flow in one direction

dioxin N− a carcinogenic chlorinated hydrocarbon occurring as an impurity in the herbicide 2,4,5-T

diploid ADJ having a basic chromosome number that is twice the number in normal gametes

disaggregate VI to separate from a mass

disc N+ a fla tened sphere

discard VT to throw away

discharge VT to give or send out

discontinuity N+ something that is not continuous

discrepancy N+ inconsistent data; the difference

disease N+ − an unhealthy condition; an illness

disintegration N− the process of breaking into small pieces

disorder N+ a disturbance of the normal working body

dispense VT to distribute or give out

displacement N+ the amount of water the load of a ship replaces

disposal N− the process of getting rid of

disrupt VT to interrupt the flow or continuity of

dissect VT to cut apart in order to examine

distill VT to purify by boiling and condensing

distillation N+ − turning a liquid into vapor

distillation column N+ an apparatus for separating petroleum into kerosene, gasoline, etc.

distorted ADJ pulled or twisted from normal shape

disturbed ADJ mentally ill

domesticate VT to bring animals under control, to tame

doughnut N+ a disc with a hole in the center, a torus

down ADJ (informal) not functioning

downstream   ADJ   in the direction in which a river flows

drain   VT   to draw off liquid by means of pipes or other apparatus

drainage   N–   drains collectively

drill   N+   a machine for boring holes

drive   N+   the device that transmits power to machinery

drop   N+   a small pear-shaped mass of liquid

drought   N+ –   continuous dry weather

dry cell   N+   a sealed battery

dry cleaning   N–   cleaning clothes with a solvent, not water

duct   N+   a tube for carrying liquid

duodenum   N+   the first part of the small intestine

dust   N–   fine particles of matter

Dutch elm disease   N–   a fungus that affects elm trees

dynamite   N–   a powerful explosive made of nitroglycerin

dysentery   N–   a disease that causes severe diarrhea

eclipse   N+   the blocking of light by the shadow of a heavenly body

ecology   N–   the study of living things in relation to each other

ectomorph   N+   a person with a light bone structure

eddy   N+   a swirling patch of water or air

effluent   N+ –   something that flows out; sewage

ejection   N+ –   a process of throwing out forcefully

elastic   ADJ   returning to original shape after stretching

electrocardiagram   N+   the pattern produced from the measurement of a heart beat

electrode   N+   either terminal of an electric source

electroplating   N–   a coating with metal by electrolysis

element   N+   a kind of atom; a part of a whole

elevator   N+   a device for transporting people in a building

elliptical   ADJ   having an oval shape

elm   N+   a deciduous tree

embryo   N+   an animal before birth

emerge   VI   to come into view, to appear

emergency   N+   a serious situation requiring prompt attention

emigrant   N+   a person who leaves one country for another

emission   N+   a substance that is sent out, e.g., pollution

emit   VT   to send out light, heat, fumes, etc.

enable   VT   to allow, to make possible

encode   VT   to put into a system of symbols for messages

endomorph   N+   a person with a heavy or fleshy build

endurance   N–   the ability to withstand strain

engage   VIT   to interlock for the transmission of power

enhancer   N+   something that improves characteristics

enlarge   VT   to make larger

ensuing   ADJ   following, describing what comes after

environment   N+ –   the surroundings

enzyme   N+   a natural protein that aids chemical processes

eon   N+   a very long time

epidermis   N+   the outer layer of the skin

episode   N+   an incident or event

epithelial   ADJ   relating to the skin

equator   N+   the imaginary line around the earth equidistant from the North and South Poles

equilibrium   N–   a state of balance

eradicate   VT   to remove permanently, to get rid of

erode   VIT   to wear away gradually

eruption   N+   the emission of lava, gas, etc., from a volcano

essential   ADJ   necessary

etch   VT   prepare a surface for painting

ethane   N–   a colorless odorless gas, C_2H_6

ethanol   N–   ethyl alcohol

ethyl alcohol   N–   the alcohol in liquor, C_2H_6O

ethylene glycol   N–   a chemical used as antifreeze, $C_2H_6O_2$

eucalyptus   N+   an evergreen tree common in Australia

eucaryote   N+   a cell with genetic material enclosed in the nucleus

eustachian tube   N+   a tube connecting the ear to the throat

evacuate   VT   to remove air or people

evaporate   VIT   to turn into vapor; to dry

evergreen   N+   a tree with leaves all year round

evidence   N–   proof, data

evolve   VI   to develop by evolution

exceed   VT   to go beyond the limit of, to be greater than

excision   N+ –   the removal of an unwanted piece by surgery

excreta   N–   waste matter expelled from the body

excrete   VT   to expel waste matter from body or tissues

execute   VT   to perform

exert   VT   to bring a quality or influence into use

exhaust   N–   the expulsion of waste gases from an engine

expel   VT   to force or send out

exposure   N–   the act of uncovering

expression   N–   the mathematical symbol for a quantity

extinct   ADJ   no longer existing in living form

extinguish   VT   to put out a fire or a flame

extract VT to obtain by means of pressure or solvent

extraterrestrial ADJ from another planet

exude VIT to flow out slowly

fabric N– cloth, woven material

facilitate VT to make easier, to aid

fail VI to not succeed

fake N+ something that looks genuine but is not

fan N+ a device for creating a current of air

fatigued ADJ weakened by repeated stress

faucet N+ a controllable outlet for water

fault N+ a break in continuity of layers of rock

faulty ADJ imperfect, having defects

feasible ADJ able to be done, possible

feed VT to supply material to; to give food to

ferry N+ a boat used to carry people, cars, etc.

fertile ADJ rich in materials needed for support

fever N+ – an abnormally high body temperature

fiberglass N– fabric made from glass fibers

fibroblast N+ a vertebrate cell that forms collagen

fibrous ADJ made of thin strands of tissue

filament N+ a very thin wire or strand

file N+ a collection of arranged items

fission N– the splitting of atoms for energy or cells for reproduction

flammable ADJ able to be set on fire easily

flare N+ an outburst of flame

flash N+ a sudden burst of light

flask N+ a type of laboratory glassware

flea N+ a small insect that feeds on blood

flexible ADJ able to bend without breaking

float VIT to rest on the surface of a liquid

flood N+ a great quantity of water covering a normally dry area

flow chart N+ a diagram of movement in a complex activity

flowage N– the act of flowing or flooding

fluctuate VI to vary irregularly

fluorescent ADJ emitting radiation while absorbing radiation from other source

fluorine N– an element (F)

flux N+ a continuous succession of changes

foliage N– the leaves of a tree or plant

forage crop N+ a crop used as food for cattle

forceps (sing.) N– pincers used for gripping by doctors or others

forecast N+ a statement that tells something in advance

forerunner N+ something that came before

formaldehyde N– a preservative chemical, CH_2O

fossil N+ the remains of an ancient species in rock

foundation N+ the strong base of a building

fraction N+ a small part of; a ratio less than one

fractional ADJ of or forming a fraction; very small

fractionating ADJ capable of breaking into parts

fracture N+ a break in a bone, rock, etc.

frame N+ a rigid supporting structure

freeway N+ a limited-access highway

freight N– goods shipped in containers

friction N– the resistance of one surface against another

frost N– a white coating of frozen vapor

fructose N– a form of sugar found in fruit

fuel N+ – material burned as a source of energy

full-term ADJ completely developed (embryo)

fume hood N+ a device for removing gases from a lab

fumigant N+ a substance used in fumigation

fumigation N– disinfection by means of fumes

fund VT to supply money for a particular purpose

fungus (pl: -gi) N+ – a nongreen plant form, e.g., mushroom

funnel N+ a conical pipe for pouring liquids into small-necked containers

furnace N+ an enclosed space for heating air or metals

fuse VI to form into a whole

fusion N– the union of atomic nuclei to release energy

galaxy N+ an independent system of stars in space

gamete N+ a sexual cell

gasket N+ a sheet of material used for sealing a joint between metal surfaces

gasohol N– gasoline mixed with alcohol

gasoline N– motor fuel distilled from petroleum

gastric ulcer N+ an open sore in the stomach lining

gear N+ a toothed wheel for transmitting power

gel N– a semisolid substance

gene N+ the chemical unit that carries heredity

generator N+ a machine that converts mechanical energy into electricity

geochemist N+ an expert in the chemistry of the earth

germ N+ a microorganism capable of causing disease

germinate VI to begin to develop and grow

gestation period N+ the period of pregnancy in mammals

glaciation N– the state or process of being covered with glaciers

glacier N+ a moving river of solid ice

gland   N+   an organ that secretes substances used by the body

glassware   N−   laboratory objects made of glass

globe   N+   the world

glucose   N−   a form of sugar, $C_6H_{12}O_6$

glue   N−   a sticky substance used for joining things

gluon   N+   a hypothetical massless particle concerned with the strong interactions between quarks

glycolysis   N−   the metabolism of sugar into energy

glycoprotein   N+ −   a complex compound of a protein and a polysaccharide

goiter   N+ −   an enlarged thyroid gland

Golgi apparatus   N+   a structure in cell cytoplasm

gorge   N+   a narrow, steep-sided valley

gradient   N+   the amount of slope in a road or railroad; a graph slope representing an increase or decrease

grain   N+   a small, hard particle; the seed of a food plant

granite   N−   hard, gray or pink rock

grant   N+   an amount of money given for a definite purpose

graphite   N−   a soft, black form of carbon used in pencils

gravitation   N−   the force of gravity or attraction between objects

graze   VT   to eat grass

grease   N−   a semisolid oily substance used for lubricating

grind   VT   to crush into powder; to sharpen by friction

grizzly bear   N+   a large bear of North America

ground station   N+   a place on the earth for transmitting or receiving radio signals from space

guidance   N−   a system for keeping an object on course

gum   N+   (dental) the flesh that surrounds the teeth

gunpowder   N−   an explosive

gynecologist   N+   a doctor of the female reproductive system

gypsum   N−   a chalklike substance, $CaSO_4 \cdot 2H_2O$

gyroscope   N+   a spinning wheel that stabilizes vehicles

habitat   N+   a place where something lives

hafnium   N−   an element (Hf)

half-life   N+   the time it takes the radioactivity of a substance to become half its original value

halide   N+   a compound containing F, Cl, Br, I, or At

halogen   N+   any of the elements F, Cl, Br, I, or At

halogenation   N−   the addition of a halogen to a compound

harvest   VT   to gather a crop

hatch   VIT   to emerge from an egg

hazardous   ADJ   dangerous, risky

heal   VI   to become healthy again

heart attack   N+ −   the sudden failure of the heart to function

heat exchanger   N+   a device for transferring heat

heliocentric   ADJ   with the sun at the center

helix   N+   a spiral

helmet   N+   a protective head covering

hemisphere   N+   half of a sphere or globe

hemodialysis   N−   the artificial cleansing of the blood

hemoglobin   N−   an oxygen-carrying substance in the blood

hemophiliac   N+   a person subject to uncontrolled bleeding

hepatitis   N−   an inflammation of the liver

herbicide   N+   a plant killer

herpes   N−   a viral disease causing blisters on the skin

herpetologist   N+   a specialist in reptiles

hibernation   N−   the passing of winter in a sleeplike state

hieroglyphic   N+   ancient Egyptian writing

hinge   N+   the joint on which a door or lid swings

hive   N+   a container for bees to live in

holistic medicine   N−   an approach to medicine concerned with the whole body, not just a particular area

hood   N+   a covering

hormone   N+   a natural body chemical with a specific duty

hull   N+   the basic frame of a boat

hurricane   N+   a storm with violent winds

hybrid   N+   an animal or plant produced from two different species

hybridization   N+ −   crossbreeding

hydraulic   ADJ   operated by the movement of fluid

hydrocarbon   N+   a compound of carbon and hydrogen in fuel

hydrofluoric acid   N−   a corrosive liquid, HF

hydrogen   N−   an element (H)

hydrolysis   N−   the chemical reaction of a compound with water, producing two or more components

hydroponic solution   N+   a nutrient liquid for growing plants without soil

hyperkalemia   N−   deficiency in extracellular potassium

hypertonic   ADJ   having a higher osmotic pressure

hypocotyl   N+   the stem of a young plant

hypothalamus   N+   the brain organ that controls temperature

hypothesis   N+   an idea about why something is true

I-beam N+ a metal girder shaped like an I

iceberg N+ a mass of ice floating on the sea

icosahedron N+ a regular polyhedron with twenty sides

ignite VIT to cause to burn; to catch fire

image N+ a picture

immerse VT to drop or dip into a liquid

immune system N+ the body's method of preventing infection

immune-deficient ADJ not resistant to infection

impact N+ the force exerted when two bodies collide

impedance N+ − the resistance of a circuit to the flow of current

impervious ADJ not able to be penetrated or influenced

implant VT to insert into a living thing

implement VT to put into effect or use

impulse N+ (med) the stimulating force in a nerve

impurity N+ − an undesirable substance in a uniform mass

incandescent ADJ glowing with heat, shining

incentive N+ −· something that encourages one to act

incinerate VT to reduce to ashes, to burn, to destroy

incision N+ a (surgical) cut made in a body

incisor N+ a sharp-edged front tooth

increment N+ a measured increase or decrease

incubator N+ a device for keeping premature young animals warm

incur VT to bring upon oneself

indicator N+ something that shows or points to

induce VT to produce, cause, or persuade

induction coil N+ a device for producing a large pulse of electricity

infection N+ − a diseased condition

infertility N− the condition of not being able to reproduce

influenza (flu) N− a viral disease of the respiratory system

inhalation N+ − an intake of breath

inhale VT to take air into the lungs

inhibit VT to restrain or prevent

injection N+ the forcing of a liquid into a body

inoculate VT to inject with a vaccine, to make immune

input N− that which is entered or put in

insert VT to put a thing between or among

insomniac ADJ unable to sleep

install VT to set an apparatus in position for use

instrumentation N− instruments collectively

insulate VT to cover with material to prevent the loss or gain of temperature

insulin N− the hormone controlling sugar absorption

intact ADJ whole, complete, not broken

integument N+ the skin

intelligibility N− the ability to be understood

intensity N+ − the quantity or rate of flow of energy

interdigitate VI to interlock or interweave

interferometer N+ a highly accurate light-measuring instrument

interferon N− a protein that stops a virus

intravenous ADJ into a vein

in vivo ADV in a living organism

iodine N− an element (I)

ionize VT to convert into ions

iridium N− an element (Ir)

irradiate VT to subject to radiation; to shine upon

irretrievable ADJ unable to be brought back

irrigation N− a system for supplying an area with water

isolate VT to place apart or alone

isomerism N− the property of having the same molecular formula but a different structural formula

isotope N+ the same element with a different atomic weight

jaw N+ the bones that form the mouth

jellyfish N+ a floating sea animal with a semisolid body

jet N+ a stream of water or air

jet lag N− the delayed effect of tiredness after flying across time zones

kaolinite N− a very pure white clay used in porcelain

keratin N− the protein in skin, nails, and hair

kerosene N− a fuel oil distilled from petroleum

kidney N+ an organ that removes waste products

kikui nut N+ an inedible nut whose oil is used as a wood preservative

killer T cell N+ a lymphocyte that passes through the thymus before killing foreign cells

kinetic ADJ of or produced by movement

knit VI to unite or grow together

knurled ADJ with projecting ridges to aid grip

label N+ a piece of paper or metal for identification

labiate ADJ like a lip or lips

lagoon N+ a saltwater lake close to the sea

larva (pl: -ae) N+ the first stage of an insect

laser N+ a device that generates intense light

lateralize VT to develop brain functions before puberty

latex N− the fluid from the cut surface of a plant

lathe N+ a machine for shaping rotating wood or metal

lattice N+ a framework of bars with spaces

launch VT to energetically send on a course

lava N− molten rock from a volcano

layer N+ a thickness of material

leach out VIT to remove by dissolving

leaf N+ a flat, usually green part of a plant

leak N+ a hole through which a substance escapes

legume N+ a plant that bears seeds in pods

lens N+ a curved piece of glass used in optical devices

lens cap N+ a cover for a lens

leukemia N− cancer of the blood

lid N+ a hinged or removable cover

lie detector N+ a device for indicating if someone is not telling the truth

life-span N+− the length of life

ligase N+ an enzyme that catalyzes a bond

lightning N− a natural flash of electricity in the sky

lignification N− the process of becoming like wood (i.e., unpalatable)

limb N+ a projecting part of a body or a tree

limestone N− rock consisting of calcium carbonate

linear accelerator N+ an accelerator that lies in a straight line

link VT to join, to connect

lipid N+ an insoluble substance in fat

liposome N+ an aqueous compartment enclosed by lipids

litmus paper N− treated paper for determining acidity

liver N+ an organ that is important in metabolism

lizard N+ a reptile with scaly skin and four legs

localization N+− the confinement to a particular space

locomotive N+ an engine for pulling a train

log N+ a length of tree trunk that has been cut

loop N+ a shape or circuit that returns to the starting point

loudspeaker N+ a device that changes electricity into sound

lubrication N− the act of supplying oil or grease

lug N+ an earlike projection usually used for attaching things

lumber N− timber (raw wood) cut into planks

lump N+ a hard mass or swelling

lymph N− the colorless fluid from tissues

lymphocyte N+ a spherical white blood cell

lyse VT to cause to undergo lysis

lysis N− cell destruction

lysozyme N+ an enzyme that attacks bacteria

machete N+ a broad, heavy knife

machine VT to make or produce by machine

machinery N− machines collectively

macrophage N+ a cell that removes foreign particles from blood

maggot N+ a larva, especially of the bluebottle fly

magma N− semimolten rock under the earth's crust

magnesium N− an element (Mg)

magnetite N− a magnetic iron oxide

mahogany N− a hard, reddish brown wood

malady N+ an illness, a complaint

malathione N− an insecticide that contains phosphorous

malfunction VI to function incorrectly

mammal N+ the class of animals that suckle young

manifest VIT to become apparent or visible; to give signs of

manometer N+ a device for measuring gas pressures

mantle N+ the area between the earth's crust and the core

manufacture VT to make in a factory

manure N− a substance used as fertilizer

map VT to make a picture of a planet's surface

margin of safety N+ the degree of safety above the essential minimum

marine ADJ of or living in the sea

marsupial N+ a mammal that carries its young in a pouch

martian ADJ concerned with the planet Mars

mate (with) VT to put two animals together to breed

mayfly N+ a flying insect

measles N− an infectious disease characterized by red spots

medium N+ a substance in which something exists

melanoma N+− a vicious form of skin cancer

membrane N+ a thin tissue that allows the controlled passage of liquids

meniscus N+ the curved surface of a liquid in a tube

mesh VI fit together, engage with

mesh current N+ the current that flows around a circuit

mesomorph N+ a person with an athletic build

metabolism N+− the process by which food becomes energy

metabolize (into) VT to transform by natural cell processes

metallurgical ADJ concerned with properties of metals

meteorite N+ a fragment reaching earth from space

meteorologist N+ one who studies the weather

methane N− a colorless, inflammable gas, CH$_4$

methyl-ethyl ketone N− a strong solvent

microbe N+ a microorganism that often causes disease

micrometer N+ an instrument for measuring small distances

microorganism N+ a microscopic plant or animal

microprocessor N+ a miniature computer

migrate VI to move from one place to another

milky ADJ not clear, cloudy

millipede N+ a small, many-legged crawling creature (Diplopoda)

mimic VT to imitate, to pretend to resemble

miniaturization N− the process of making very small

minnow N+ a small freshwater fish

missile N+ a weapon for projecting at a target

mission N+ an organized effort for a definite purpose

mitochondria (pl) N+ the parts of a cell that release energy

mitosis N− the division of a cell or a nucleus

module N+ a standardized part, an independent unit

moisture N− water vapor

molar N+ any of the teeth at the back of the mouth

mold N+ − tiny fungi that form in warm places

mole N+ 6.02×10^{23} atoms

molecule N+ a group of atoms bound together

molten ADJ liquefied for a period of time

molybdenum N− an element (Mo)

moment N+ the tendency to produce motion around a point

momentum N+ − the quantity of energy gained by movement

monitor VT a person or device that observes

moraine N+ a mass of rock carried by a glacier

morphine N− a pain-relieving drug made from opium

mortality N− the death rate

mount VT to put into place on a support

mucous membrane N+ the moist epithelium inside the gut

mud N− soft, wet earth

muffler N+ a device that reduces the noise of an engine

multimeter N+ a circuit analyzer that measures volts, ohms, etc.

mumps N− a viral disease that causes swelling in the parotid glands

muscle N+ − animal tissue that produces movement

musk N− a secretion of the musk deer used in perfume

mutagenic ADJ capable of changing genetic structure

nail N+ a small metal spike for joining wood

nanosecond N+ 1×10^{-9} seconds

narcotic ADJ causing sleep

nasal ADJ of the nose

natality N− the birthrate

nauseated ADJ having a feeling of sickness

navigation N− the act of directing the course of a ship or plane

nebula N+ a large celestial structure composed of gas

nectar N− a sweet fluid produced by plants

needle N+ a long thin piece of metal

negligible ADJ of a small amount not worth measuring

neon N− an element (Ne)

neurotransmitter N+ a substance that transmits nerve impulses

neutralize VT to balance by means of an opposing effect

neutron star N+ the invisible final stage of a star

nickel N− an element (Ni)

nicotine N− a poisonous substance found in tobacco

niobium N− an element (Nb)

nitric acid N− a corrosive acid, HNO$_3$

nitrocellulose N− a cellulose nitrate used in lacquers

node N+ a point of attachment

nodule N+ a small, rounded lump; an irregularly shaped rock

nostril N+ either of the two openings of the nose

notch N+ a V-shaped cut or indentation

noxious ADJ unpleasant and harmful

nucleophilic ADJ attaching to a compound where electrons are needed

nucleus N+ the center of a cell, atom, etc.

nurse N+ a person trained to care for the sick

nutrition N− the process of providing adequate food

objective ADJ having existence outside a person's mind

obstacle N+ a thing that blocks progress

olefin N+ an open-chain hydrocarbon with double bonds

opaque ADJ not clear, not transparent

optical fiber N+ − glass fiber used to transmit light

optometrist N+ an eye doctor who also fits glasses

orbit N+ the curved path of a planet, satellite, etc.

orchid N+ a plant with exotic flowers

order N+ a classification of similar plants or animals

ore N+ rock containing a metal in some form

organ   N+   a part of an animal or plant with a definite function

organism   N+   a living animal or plant

oscilloscope   N+   an instrument for showing visually the changes in a varying current

osmium   N−   an element (Os)

ostrich   N+   a large, swift-running bird that cannot fly

outlet   N+   a source of electricity, gas, water, etc.

output   N+ −   the amount produced

oval   ADJ   egg-shaped

overflow   N−   that which spills if a container is full

oxidation   N−   the process of combining with oxygen

oxygen   N−   an element (O)

ozone   N−   a form of oxygen with a sharp smell, O_3

p-level   N+   the energy required to change a p-orbital

p-orbital   N+   an electron orbit with a dumb-bell shape

pair   VT   to put two things together

palatable   ADJ   tasty, capable of being eaten

paleontology   N−   the study of fossils

panda   N+   a Chinese black and white bear-like mammal

papillary   ADJ   of the protuberances on the skin or tongue

parabola   N+   a curve like that made by a thrown object

paradox   N+   a statement that seems to contradict itself

paramecium (pl: -a)   N+   a one-celled animal with cilia

parameter   N+   a quality that restricts or gives order to

parasite   N+   an animal or plant that lives off another

parotid gland   N+   the salivary gland below the ear

particle   N+   a very small portion of matter

particulate   N+   a minute separate particle

patient   N+   a person receiving treatment from doctor

peak   N+   the highest point

pebbled   ADJ   having the appearance of rounded stones

pedal   N+   a lever operated by the foot

peg   N+   a wooden or metal pin for fastening

pelican   N+   a large water bird with a pouch in its bill

pellet   N+   a small, rounded, closely packed mass

penetrate   VT   to pierce; to enter and permeate

penicillin   N−   an antibiotic from mold fungi

peptide unit   N+   a combination of amino acids in which an amino group is united with a carboxyl group

perforation   N+   a small hole in a sheet of metal or paper

periodic table   N+   a regular chart of the elements

peritoneum   N+   the membrane lining the abdomen

permeabilize   VT   to make something permeable

personnel   N−   the people who work in a place

perturbation   N+   a disturbance; a cause of anxiety

pest   N+   an animal that destroys crops or other growing things

pesticide   N+   a substance for destroying harmful insects

petal   N+   a colored outer part of a flower head

petri dish   N+   a small, covered, shallow, glass lab dish

petrified wood   N−   wood that has been converted to stone

petroleum   N−   mineral oil found underground

phage   N+   an alternative term for **bacteriophage**

pharmaceutical   N+   a medicinal drug

pharmacy   N+   a place that prepares and sells drugs

phase   N+   a stage of change or development

phenol   N−   an organic antiseptic, C_6H_6O

pheromone   N+   a secretion used by animals for identification

phloem   N−   the soft tissue of plant stems

phosphate   N+   a salt or ester of phosphoric acid

phosphodiester   N+   a compound containing esters and phosphorous

photon   N+   a quantum of light energy

photosynthesis   N−   the plant process for making energy

phototropism   N−   the movement towards light

photovoltaic   ADJ   of a photoelectric cell in which light causes the generation of electricity

pickle   N+   a vegetable preserved in vinegar

pie diagram   N+   a circular chart sectioned into wedges that indicate percentages

pier   N+   a dock built out into the sea

pile   N+ −   the thickness of a carpet

pilot light   N+   a small jet of gas for lighting a burner

pine   N+   an evergreen tree with needle-shaped leaves

pineal gland   N+   a cone-shaped organ in the brain

pipette   N+   an instrument for measuring a small quantity

piston   N+   a sliding cylinder in a tube

pit   N+   a depression in the skin or a surface

plane   N+   an imaginary flat or level surface

plasma   N+   a kind of gas with an equal number of positively and negatively charged particles

plate   N+   a thin, flat structure

platform N+ a level surface raised above the ground or sea

pliers N+ a pincerlike tool for gripping things

plow VT a farm instrument for turning soil

plug N+ a device with metal pins for making an electrical connection

plunger N+ a part of a mechanism that moves a liquid

plutonium N− an element (Pu)

pneumonia N− a disease of the lungs

poisonous ADJ having the effect of destroying life

polar icecap N+ the mantle of ice around the North Pole

polarity N+ − the property of having positive and negative poles

pole N+ a positive or negative terminal or axis

polish VT to make smooth by rubbing

pollen N− the powdery substance by which flowers reproduce

pollination N− the fertilization of flowers

pollutant N+ something that makes unclean

polonium N− an element (Po)

polyallomer N+ a substance with a varying chemical constitution but the same crystalline form

polyhedron N+ a solid figure with seven or more sides

polymerase N+ an enzyme that links many molecules

polyp N+ a simple organism with a tubelike body

polypeptide N+ a peptide with more than two amino acids

polysaccharide N+ a complex carbohydrate or sugar

porcupine N+ a rodent whose body is covered with spines

pore N+ a tiny opening in a skin or a leaf

porpoise N+ a sea animal with a blunt snout

potable ADJ drinkable

potent ADJ able to have a strong effect

precede VT to come before

precipitate N+ a substance deposited from a solution

precipitation N− rain, snow, or hail; the process of producing a solid from a solution

predator N+ an animal that hunts and eats others

premature ADJ occurring before the proper time

prestressed ADJ (of concrete) strengthened with stretched wires within

prey N− an animal killed by another for food

primordial ADJ of the earliest times of the world

printout N+ computer content produced in printed form

probe N+ a device or test for intense investigation

proboscis N+ a large animal nose

profile N+ a side view

prognosis N+ a forecast of the course of a disease or other condition

projectile N+ something thrown forcefully

proliferation N+ − the rapid production of new growth or offspring

prominence N+ a cloud of incandescent gas from the sun

propane N− a gaseous hydrocarbon, C_3H_8

propeller N+ a revolving device for making ships move

propulsion N− the process of being pushed forward

protein N+ − an organic compound containing nitrogen

protocol N+ a correct or established procedure

puberty N− the stage at which a person is first capable of producing offspring: ages twelve to fourteen

pulley N+ a wheel over which chain or rope moves

pulp N− a moist mass of material; the wood basis of paper

pulsar N+ a cosmic source of pulsating radio signals

pulse N+ the rhythmic throbbing of arteries as blood is pumped through them

pump N+ a device for moving a liquid

pupation N− the stage of insect life after the larva

purify VT to cleanse, make pure

pyrethrum N− a powder for exterminating insects

Pyrex glass N− heat-resistant glass

pyrimidine N+ a source of nitrogenous bases that make up nucleotides, $C_5H_5N_2$

python N+ a kind of large snake

quadrant N+ any one of four parts of an area

quantum (pl:-a) N+ a unit quantity of energy proportional to the frequency of radiation

quark N+ one of three components of elementary particles that form matter

quasar N+ a starlike object with high radiation

query VT to question or survey

quill N+ a porcupine's defensive spine

quotient N+ the result of division

rabies N− a viral disease of the central nervous system that occurs chiefly in carnivorous animals

radar N− a system for detecting the presence of objects with radio waves

radical ADJ fundamental; drastic; thorough

radicle N+ the root of an embryo in a developed seed

radium N− an element (Ra)

radon N− an inert element (Rn)

random ADJ without a pattern, principle, or purpose

ranges N+ the limits between which something varies

rare earths N+ a group of metallic elements

reactor N+ an apparatus for the controlled production of nuclear energy

reagent N+ a substance that produces a chemical action

rear up VI to raise up on the hind legs

receptor N+ a device for receiving a signal

reclamation N− the process of making usable again, especially land

recombinant ADJ of genetic material in new combinations

recurrence N+ something that happens again

redwood N+ a California conifer with reddish wood

reef N+ a ridge of rock or coral near the ocean surface

refine VT to remove impurities from

reflector N+ a thing that throws back light or images

refraction N− the bending of light at the boundary of two mediums

regenerate VT to restore by growing new tissue

region N+ an area with definite characteristics

regulate VT to control by rules and restrictions

relative ADJ considered in relation to something else

relay N+ a device that activates changes in a circuit

remission N− a reduction of force or intensity

replica N+ an exact copy or reproduction

replicate VT to reproduce; to repeat the same procedure

repress VT to prevent emotions from finding an outlet

reproduce VIT to produce offspring; to copy

research N− an investigation

resemble VT to appear similar to, to look like

reservoir N+ a lake that stores water; a container that supplies fuel or other liquids

residue N+ the substance left after combustion or evaporation

resin N− a sticky substance from pine trees

resistance N+ − the ability to prevent or oppose

respiratory ADJ involving breathing or oxygen absorption

restriction N+ − a limit; a control

retard VT to cause to go slower

reticular ADJ like a net in operation or effect

retina N+ the light-sensitive area of the eye

revive VT to bring back to life or consciousness

rheostat N+ a device that varies current

rhodopsin N− the purple pigment in the eyes that is involved in sight

Richter scale N+ a scale for measuring earthquakes

ridge N+ a narrow, raised strip

rift zone N+ a volcanic area of splitting or separation

rig N+ equipment for a special purpose

rigid ADJ stiff, not bending or flexible

rim N+ the edge of something circular

ripen VI to become mature or ripe

rivet N+ a metal connecting device

roadbed N+ the material laid down for a road

rod N+ a slender, straight, round stick or metal bar

rodent N+ a small animal with strong front teeth

roller bearing N+ a bearing in which a shaft rotates in contact with rollers in a cage

root N+ the part of a plant in the soil

rope N+ − a strong, thick cord

roughly ADV very approximately

round off VT to approximate to the nearest whole number

rub VT to press against a surface and slide

rubber N− a tough, elastic substance made from plants

rugged ADJ rough and unyielding; having an uneven surface

rupture VT to burst or break

rust N− iron oxide

saline ADJ containing salt or salts

salmon N+ a large fish with pinkish flesh

salverform ADJ tubular with a spreading expansion at one end

sample N+ a small quantity for testing

sandblasting N− cleaning or treating with a jet of sand

sandstone N− rock formed of compressed sand

sanitation N− an arrangement to protect public health by the efficient disposal of sewage

saturated ADJ caused to absorb as much as possible

sawdust N− powder produced by the sawing of wood

scale N+ an ordered series of units

scalpel N+ a doctor's knife

scan VT to sweep a (radar) beam over to find something

scar N+ the mark left where a wound has healed

scarce ADJ not enough to supply a demand; rare

scent N+ the characteristic smell of something

schematic diagram N+ a chart showing the plan or structure of something

schizophrenia N− a mental disorder in which a person acts irrationally

scorpion N+ a small animal with claws and a sting (arachnid)

screen VT to examine systematically; to control the entrance

screw N+ a metal pin with a spiral ridge for fastening

screwdriver N+ a hand tool for turning screws

scrub VT to rub hard with something coarse; to clean

sea anemone N+ a tube-shaped sea animal

seal VT to close securely to prevent penetration

seam N+ the line where two edges join

sebaceous ADJ of certain skin glands that secrete oils

secrete VT to send out a substance into the body

sediment N+ − fine particles of matter that settle

seed VT to place particles in a cloud in order to cause condensation

seep VI to ooze slowly through

seethe VI to bubble or surge, as in boiling

seismograph N+ an instrument for recording earthquakes

seismometer N+ an instrument that shows the force or direction of an earthquake

selective ADJ choosing carefully

selenium N− an element (Se)

semiconductor N+ a substance that conducts electricity but not as well as a metal

senility N− a mental weakness from old age

sensor N+ something that senses

serosa N+ the serous membrane lining the pleural, peritoneal, and pericardial cavities

serotonin N− a neurotransmitter, 5-hydroxy-tryptamine

serum N− the fluid that remains when blood is clotted

session N+ a meeting for deciding something

sewage N− waste matter drained from houses

sewer N+ the drain for carrying away sewage

shaft N+ a long metal rod for transmitting power; the vertical passage of an elevator

shale N− oily stone that splits easily

shear strength N− the ability to withstand lateral shifting

shipworm N+ a worm-shaped mollusk that bores into ships

shiver N+ a trembling movement from cold or fear

shock N− the effect of a violent impact

shoot N+ a young plant

shoring N− the timbers that prevent sinking or sagging

short (-circuit) N+ a faulty electrical connection in which a current flows by a shorter path than usual

shrinkage N− the amount that something becomes smaller

shrivel VI to shrink and wrinkle from heat or cold

shroud N+ the sheet in which a dead body is wrapped

shutdown N+ the cessation of work or business

shutter N+ a device that opens and closes to admit light into a camera aperture

shuttle N+ transportation back and forth

sickle-cell anemia N− severe hereditary anemia characterized by bent cells

silicate N+ a compound containing silicon (Si)

silt N− a sediment deposited by water

site N+ the ground or place where something occurs

skull N+ the bony framework of the head, cranium

slash VT to make a sweeping cut

sleep apnea N− a life-threatening disorder in which breathing is frequently blocked

sleeping pill N+ a quantity of sleep-inducing medication

slide N+ a small glass plate on which microscope specimens are mounted

slip VI to slide accidentally

slope N+ a line that lies at a nonhorizontal angle

sludge N− a heavy, slimy deposit or sediment

smelt VT to melt ore in order to extract metal

smog N− fog polluted by smoke

smooth ADJ even, flat

snout N+ an animal nose

snow line N+ the level above which snow lies permanently

sniff VT to inhale abruptly through the nose

snuff N− powdered tobacco for sniffing

socket N+ a device for receiving an electrical connection

sodium N− an element (Na)

software N− computer programs

soil N+ − the loose layer of earth in which plants grow

solar pond N+ a small lake with layers of salt and fresh water for generating solar electricity

solar wind N+ the flow of charged particles from the sun

solder N− a soft alloy for cementing metal parts

soldering iron N+ a device for heating solder

solid-state ADJ using transistors instead of vacuum tubes

solvent N+ a liquid used for dissolving something

sonic boom N+ the noise created when a plane flies faster than sound

sort VT to arrange in order

sound ADJ solid, well-built

sour VI to become sour, to rot or decay

source N+ the place from which something comes

sow VT to put seed into the ground

span N+ length in time or distance between points

species N+ a subgroup of plants or animals in a genus

specification N+ a detailed description of how something is made

specificity N− a quality or state of being peculiarly adapted to a purpose

specimen N+ a part or an individual taken as an example

speck N+ a very small piece

spectroscopy N− the study of spectra

spectrum (pl: -a) N+ the band of colors produced by the dispersion of light

sperm N+ the male reproductive cell

spider N+ an animal with a segmented body and eight legs (arachnid)

spin VIT to turn or cause to turn rapidly on an axis

spinal cord N+ the mass of nerve fibers in the backbone

spine N+ the backbone

spleen N+ an organ that defends the bloodstream against pathogens

splinter N+ a thin, sharp piece broken from wood or stone

split VT to divide into parts

spoilage N− damage that makes something useless

sponge N+ a water animal whose porous structure is used for wiping and cleaning

spool N+ a reel on which something is wound

spot VT to detect or recognize

spread VIT to open out, to extend; to become widely suffered

spring N+ a coiled metal device; a source of underground water

sprinkler N+ a device for watering growing plants

sprout VI to begin to grow or appear; to put forth shoots

squamous ADJ like armor or scales

stable ADJ firmly established; not fluctuating

stage N+ a period in the course of development

stain N+ a discoloration; a permanent mark

stand N+ a metal device for holding instruments

staphylococcus N+ parasitic bacteria that grow on mucous membranes

starch N− a white carbohydrate, $C_5H_{10}O_5$

stationary ADJ not able to move

steam N− gaseous water

stellar ADJ concerned with stars

stethoscope N+ an instrument for listening to body sounds

stoma (pl: -ata) N+ the opening in a leaf for respiration

stomach N+ the organ for the first major part of digestion

stopper N+ a plug for closing a bottle or flask

storm N+ a violent disturbance of the atmosphere

strand N+ a single thread or strip of fiber

stratum (pl: -a) N+ one of a series of layers

stress N− pressure, tension, strain

strike VIT to hit

stroke N+ an attack of apoplexy or paralysis

studio N+ a soundproof room for recording

stunted ADJ hindered in growth or development

subatomic particle N+ a particle of which an atom is constructed

subcutaneous ADJ beneath the skin

subduction zone N+ an area where one continental plate slides under another

subjective ADJ depending on personal view

submucosa N− the connective tissue below the lining of the gut

substrate N+ an underlying layer; a substance that is acted upon

substructure N+ the groundwork; the foundation

suffocate VT to kill or die by stopping breathing

sulphuric acid N− an acid, H_2SO_4

superconductor N+ a metal with a low resistance at cold temperatures

superfluid N+ an unusual liquid form (e.g., helium II)

superstructure N+ a structure that rests on something else

supplement N+ a thing added to make up a deficiency

suppress VT to keep from being shown, seen, or known

surface N+ the outside or uppermost area

surgeon N+ a doctor who performs operations

survey N+ a general examination of land or opinions

suspend VT to hold in solution

suspension N+ a means of support; solids held in a liquid

sustain VT to suffer; to support or keep alive

swamp N+ spongy land saturated with water

sweat N− body moisture given off through the skin

swell VIT to become larger from within

swift ADJ quick, rapid

swine flu N− a type of influenza virus

switch off VT to disconnect a power source

symposium N+ a meeting for the discussion of a subject

symptom N+ a sign of the existence of a condition

synapse N+ the place where nerve cells join

syndrome N+ a number of symptoms occurring together

synthesize VT to combine separate parts into a whole; to make

synthetic ADJ made by humans

syringe N+ a device to draw in liquid for injection

T-4 ADJ referring to a type of bacteriophage

tank N+ a large container for a liquid

tanker N+ a ship designed to carry liquids

tap N+ a device for controlling liquid flow

tectonic plate N+ a structural plate in the earth's crust

telepathically ADV without speech or writing; mind to mind

temperate ADJ having a mild climate

tenant N+ a person who occupies land or a building

tensile strength N+ − the greatest longitudinal stress a substance bears

tension N− an effect produced by forces pulling against each other

tentacle N+ a part of some animals that extends for feeding

terminal N+ a point of connection in a device or circuit

thaw VIT to pass or cause to pass into a liquid or unfrozen state

theodolite N+ a surveying instrument for measuring angles

thermocouple N+ the union of two conductors to produce an electrical current

thermonuclear ADJ concerned with the changes in a nucleus that require high heat, as in a hydrogen bomb

thimble N+ a small metal cap to protect the finger in sewing; a part shaped like such a cap

thread N+ a thin length of cotton; the ridge on a screw

throat N+ the passage in the neck through which food passes

throttle N+ a valve controlling the flow of fuel or other material

thrust N+ − the quantity of propulsion

thunder N− the loud noise that accompanies lightning

thunderhead N+ a rounded cumulus cloud preceding thunder

thymidine N− a nucleotide precursor of DNA

thymine N− one of the major pyrimidines, $C_5H_6N_2O_2$

thymine dimer N+ a covalent linkage of two thymine molecules that blocks the replication of DNA

thyroid N+ a gland in the neck that secretes hormones

tidal wave N+ a very large destructive ocean wave

tide N+ the regular rise and fall in the level of the sea

tile N+ a thin slab of ceramic for floors, roofs, etc.

tire N+ a rubber tube or covering fixed around a wheel

tissue N+ − the substance forming an animal or plant body

titrimeter N+ an instrument for determining the pH of a liquid

topical ADJ designed for use on the skin, not eaten

tornado N+ a destructive, funnel-shaped storm

tortoise N+ a land turtle of warm climates

torus N+ a shape like a fat ring or doughnut

tourniquet N+ a strip of cloth drawn tightly around a limb to stop the flow of blood

toxic ADJ poisonous

toxin N+ a natural metabolic poison

trace N+ a very small quantity

track N+ marks left by moving animals; a path, a road

traffic N− vehicles or people moving on a route

tranquilizer N+ a drug to relieve anxiety

transistor N+ a semiconductor device with three electrodes

transmission N+ the gears by which power is passed to the wheels of a vehicle

transparent ADJ allowing light to pass through

transplant VT to remove and reestablish in another place

trap VT to catch or find and remove

treatment N+ − a method of cure

tree ring N+ one of the annual circles in tree trunks

trigger VT to start, to initiate; to be the cause of

trim N− light woodwork in the finish of a room, building, etc.

trimaran N+ a three-hulled boat

trunk N+ the main stem of a tree

trypsin N− a digestive enzyme in the pancreatic juice

tsunami N+ a series of tidal waves

tumble VI to roll over in a disorderly way

tumor N+ an abnormal mass of new tissue in the body

tungsten N− an element (W)

turbine N+ a motor driven by the pressure of air or water

turbulence N− violent uneven movement

twist VT to give a spiral form to

typhoid N− a bacterial disease of the intestines

ultrasound N− vibrations or waves with a frequency higher than that of audible sound

ultraviolet ADJ referring to a kind of electromagnetic radiation

unconscious ADJ not aware or awake

undergo   VT   to experience, to be subject to, to suffer

underpinning   N+   the material and construction used for support

unimpeded   ADJ   not hindered or blocked

unpalatable   ADJ   having a bad taste

uplift   N−   the process of being raised by air

upstream   ADJ   towards the source of a river

utility   N+   a public service providing gas, water, etc.

V-belt   N+   a continuous driving belt with a V shape

vaccine   N+   a preparation designed to give immunity

vacuum   N+   the space in a container without air

vacuum advance   N+   a device in an automobile engine that adjusts the position of the distributor in acceleration

validity   N+ −   the state of having logic or legality

valley   N+   a long, low area between hills

valve   N+   a device for controlling the flow in a pipe

vaporize   VIT   to become or cause to become a vapor or gas

varnish   N−   a liquid that, when dry, protects wood

vat   N+   a tank or other large vessel for holding liquids

vector   N+   something with magnitude and direction; a carrier of disease or infection

vegetation   N−   plants collectively

vein   N+   a tube that carries blood to the heart

velocity   N+ −   speed, especially in a given direction

vent   N+   a place from which gas is released

vent microbe   N+   a microorganism living near a deep-ocean vent

ventilation   N−   the circulation of air

ventricle   N+   the heart chamber that pumps blood into the arteries

venturi meter   N+   a device for measuring liquid flow in pipes

verge   N+   an edge; the point at which something begins

verify   VT   to check the truth or correctness of something

vertebrate   N+   an animal with a backbone

vessel   N+   a container; a tubelike structure in animals

vial   N+   a small bottle, especially for medicine

vinyl   N−   a kind of plastic, polyvinylchloride

viscosity   N+   the degree of resistance to flow; thickness

vitamin   N+   a chemical needed for metabolism

volatility   N+ −   ability to evaporate

volcano   N+   a mountain from which lava is expelled

voltage   N−   electromotive force expressed in volts

voltage spike   N+   a sudden sharp increase in voltage that can harm electronic components

voltmeter   N+   an instrument that measures electric potential

vortex   N+   a whirling mass of water or air

vulcanize   VT   to heat with sulfur

ward   N+   a hospital area with beds for a certain group

ward off   VT   to keep at a distance; to prevent from harming

warehouse   N+   a building for storing goods

warrant   VI   to justify; to authorize

washing soda   N+   a powder for washing, Na_2CO_3

waste   N−   a by-product of industrial or bodily processes

waterproof   ADJ   unable to be penetrated by water

waterwheel   N+   a wheel turned by the flow of water

weaponry   N−   weapons collectively

weatherproofed   ADJ   made unable to be damaged by weather

web   N+   a network of fine strands made by a spider

weed   N+   a wild plant growing where it is not wanted

weld   VT   to unite pieces of metal after softening with heat

whetstone   N+   a shaped stone used for sharpening tools

willow   N+   a tree with flexible branches growing near water

wind bracing   N−   structural support that resists wind

wind tunnel   N+   a tunnel for producing a stream of air at a known velocity for testing models

wiring   N−   all the wires in a building or apparatus

withstand   VT   to endure successfully

wolf   N+   a wild animal of the dog family

woolly mammoth   N+   an extinct elephant with long, shaggy hair

worm   N+   a tubelike animal with no backbone

X ray   N+   a kind of electromagnetic radiation

xylem   N−   woody tissue of a plant that conveys water

young   N+   children, offspring

zirconium   N−   an element (Zr)

zone   N+   an area of activity

zoology   N−   the scientific study of animals

zygote   N+   a cell formed by the union of two gametes

INDEX

multiplication, 122
subtraction, 122
may, 269, 272
may be able to, 273
means of transportation, article with, 253
meanwhile, 162
might, 272
might be able to, 273
minus, 122
modal, 269–274
  capability, 273–274: negative, 273; paraphrase,
    273; past form, 273; past negative, 273
  obligation, 269–271: negative, 270; paraphrase,
    270; past form, 270; past negative, 270; sub-
    junctive, 271
  probability, 272–273: negative, 272; paraphrase,
    272; past form, 272; past negative, 272
  words implying, 271
more importantly, 148
more than, 261
moreover, 145, 148
mostly, 103
most (of), 103
most of all, 104
much as, 145
must, 269, 272

naming verbs, 253
near, 219
negative noun comparison, 263–264
neither/nor (*see* parallelism)
nevertheless, 153
next to, 219
no less, 250
noncount noun (*see* noun, uncountable)
nonrestrictive (*see* relative clause, nondefining)
nonetheless, 153
not far from, 219
not nearly as, 261
not only . . . but (also) (*see* parallelism)
not only that, 145
not to mention, 146
not to mention the fact that, 145
notice, 115
notwithstanding, 153
notwithstanding the fact that, 153
noun, 3–10
  countable, 3–5: plural, 9; singular, 9
  dual, 5–7
  uncountable, 3–7, 8–10: mass, 4–5
noun clause, 47–56, 69
  embedded question, 47–51
  with *that,* 69, 254–257
noun compound, 264–268
  classification, 266–267
  interpretation, 265
noun phrase,
  structure, 68–69: embedded question, 69; ger-
    und, 69; noun clause with *that,* 69; verb$_{ing}$
    clause, 214

noun phrase parallelism (*see* parallelism)
nowhere near, 219

object, 18, 60
  in passive, 60
  in relative clause (*see* relative clause)
  of sentence, 18
observe, 115
off, 220, 222
on, 107, 220, 222
one, 36–37, 148
on the one hand, 152
on the other hand, 149, 152
on top of, 220
opposed to, 260
other(s), 171
otherwise, 147
out of, 223
outside, 221
over, 178, 220, 222
owing to, 157

parallelism, 68–70, 117–119
  coordinating conjunction, 69, 118: *and,* 69, 118;
    both . . . and, 69, 118; *either/or,* 69, 118; *neither/*
    nor, 69, 118; *not,* 69, 118; *not only . . . but (also),*
    69, 118; *or,* 69, 118
  noun phrase, 68–70
  verb phrase, 117–119
participial phrase, 205–215
  dangling modifier, 211–214
  sequential actions, 208–210
  simultaneous actions, 206–208
  verb$_{ing}$ clause as noun phrase
  verb$_{ing}$ clause of result, 210–211
passive voice, 59–68, 255–257
  by-agent, 63–64
  choosing active or passive, 62–64
  how-agent, 64–68; *by,* 65; *by means of,* 65–66;
    describing methods, 67–68; *with,* 65
  number, 60
structure, 59–60
tense, 60–61: with *being,* 61
past, 222
past perfect, 110–111, 175
perception verb, 115
  continuous form, 115
phrase, 143
phrase subordinator, 144
pie diagram, 186
plan of development sentence, 77, 128, 190
plus, 122
preposition, 105–109, 173–180, 218–225
  as vs. *like,* 224–225
  at, 105, 106: location that implies function, 106;
    position of time, place, or measure, 106
  in, 107: containment, 108; mode: measurement,
    direction, action, 107–108
  on, 107: surface or line, 107

preposition *(cont.)*
 place, 218–225: passage or direction, 221–224; relative position, 218–221
 time, 107, 173–180: end point or limit, 176–178; range, 173–174; starting point, 174–176; time period, 178–179
present perfect, 174, 180–183, 228–229
 vs. simple past, 180–181
prewriting activity, 23
 describing a diagram, 23–26
 describing cycles and steps, 124–125
 determining classifications, 188–189
 determining parts, 75–76
 making a financial decision, 292–294
 organizing information, 278–279
 summarizing, 239
product, 122
progressive *(see* tenses, continuous)
provided (that), 147
put more simply, 146

quotient, 122

rank, 253
ranking adjective, 36–37
 sequential, 36
 superlative, 36
 unique, 36–37
relative clause:
 condition or circumstance, 16, 41–42
 defining, 13–17, 41–47, 51–57: main clause, 41; object form (∅ form), 15–17; relative pronouns, 96; subject form (S form), 15–17; subordinate clause, 41; with preposition, 42–44
 nondefining, 94–97: object form (∅ form), 96–97; relative pronouns, 96; subject form (S form), 96–97
 reduction, 51–57, 97–103, 264–265: defining, 51–57; nondefning, 97–103; noun compound, 264–265
 relative adverb, 44–47: equivalent of *when*, 47; equivalent of *where*, 46; *when*, 45; *where*, 44–45
 relative pronoun, 13, 14, 41–43, 94, 96, 98: equivalent of *whose*, 46; *that*, 13, 41–42, 94, 96; *that vs. which*, 96; *that vs. who*, 14; *which*, 42–43, 94, 96, 98; *who*, 14, 96; *who vs. whom*, 14; *whom*, 14, 96; *whose*, 44
 result or principle, 16, 41–42
remainder, 122
report, 255
research report, 278–291
 model, 283, 291
 structure, 279–283: abstract, 239–241, 279, 283, 299, 301; acknowledgments, 282; discussion, 282; introduction, 279–280; materials and methods, 280–281; references, 283, 289–290
 results, 281–282
 topics, 291

resemblance, 258, 259
resemble, 259
restrictive *(see* relative clause, defining)
result clause, 210–211
 in this manner, 210
 so, 210
 thereby, 210
 thus, 210
reveal, 255

same, 36, 258
second, 148
see, 115
sentence,
 fragment, 19
 typical structure, 18
sequence of tenses, 227–230
 with facts, 229–230
 with personal ideas, 230
shall, 269
shared knowledge, 37–40
 cultural, 38–39
 regional/local, 39–40
 plain sight, 39*fn*
 world, 38
should, 269, 272
should be able to, 273
show, 255
similar, 259
similarity, 258, 259
similarly, 145
simultaneously, 162
since, 162, 174–176, 228–229
since (because), 156
since that time, 161
since then, 161
single event, 55, 102, 147*fn*
so, 156–157, 210, 250
some, stressed, 140
some, unstressed, 84, 134, 140–141
some/any other, 171
some of/any of the others, 171
so that, 158
so . . . that, 157
still, 153
square toot, 123
subject,
 bare, 20–21
 of sentence, 18
 passive, 59–63
 relative clause *(see* relative clause)
subjunctive, 271
subordination, 143–165
subordinating adverb, 143
subsequently, 161
such as, 148, 225
such . . . that, 157
sum, 157
surrounded by, 218